TQM
AN INTEGRATED APPROACH

Implementing Total Quality through Japanese 5-S and ISO 9000

SAMUEL K HO

KOGAN PAGE

YOURS TO HAVE AND TO HOLD
BUT NOT TO COPY

The publication you are reading is protected by copyright law. This means that the publisher could take you and your employer to court and claim heavy legal damages if you make unauthorised photocopies from these pages. Photocopying copyright material without permission is no different from stealing a magazine from a newsagent, only it doesn't seem like theft.

The Copyright Licensing Agency (CLA) is an organisation which issues licences to bring photocopying within the law. It has designed licensing services to cover all kinds of special needs in business, education and government.

If you take photocopies from books, magazines and periodicals at work your employer should be licensed with CLA. Make sure you are protected by a photocopying licence.

The Copyright Licensing Agency Limited, 90 Tottenham Court Road, London, W1P OLP. Tel: 0171 436 5931. Fax: 0171 436 3986.

First published in 1995

Apart from any fair dealing for purposes of research or private study, or criticism or review, as permitted under the Copyright, Designs and Patents Act, 1988, this publication may only be reproduced, stored or transmitted, in any form or by any means, with the prior permission in writing of the publishers, or in the case of reprographic reproduction in accordance with the terms and licences issued by the CLA. Enquiries concerning reproduction outside those terms should be sent to the publishers at the undermentioned address:

Kogan Page Limited
120 Pentonville Road
London N1 9JN

© Samuel K Ho, 1995

British Library Cataloguing in Publication Data

A CIP record for this book is available from the British Library.

ISBN 0 7494 1561 4

Typeset from the author's disk
Printed and bound in Great Britain by Biddles Ltd, Guildford and King's Lynn

CONTENT iii

FOREWORDS viii

PREFACE x

ABBREVIATIONS xii

1 Introduction to TQM 1
    The Japanese Quality Miracle 1
    What is TQM? 4
    Relationship between TQM and Corporate Strategy 6
    Some Important TQM Concepts 11
    Why is there a Need for TQM? 13
        Chapter Summary 17
        Self-Assessment 1.1: Quality Consciousness Checklist 17
        Self-Assessment 1.2: TQM Consciousness Checklist 18
        Self-Assessment 1.3: The F-Test 19
        Hands-on Exercise 19

2 The TQM Gurus' Ideas 21
    Introduction: Who are the TQM Gurus? 21
    W. Edwards Deming 22
    Joseph M. Juran 30
    Philip Crosby 32
    Kaoru Ishikawa 35
    Shigeo Shingo 36
    Yoshio Kondo 38
    Why are TQM Gurus' Ideas Useful? 40
    How to Implement TQM Gurus' Ideas 41
        Chapter Summary 42
        Hands-on Exercise 42

3 The TQMEX Model — 43
- Introduction: The Need for a Model in TQM — 43
- What is TQMEX? — 44
- Is TQMEX Logical? — 49
- Pre-requisites for Implementing TQMEX — 50
- The 4Cs of TQM — 52
  - Chapter Summary — 60
  - Self-Assessment Exercise 3.1: Rate your Commitment — 60
  - Self-Assessment Exercise 3.2: Prioritise your Competence — 61
  - Self-Assessment Exercise 3.3: Rate Your Communication — 61
  - Hands-on Exercise — 62

4 Japanese 5-S Practice — 63
- Introduction: What is the 5-S? — 63
- The 5-S in Details — 64
- The 5-S in the Office — 79
- Why is the 5-S Useful? — 82
- How to Implement the 5-S — 84
  - Chapter Summary — 91
  - Hands-on Exercise — 91
  - Annex 4.1: 5-S Audit Worksheet — 92
  - Annex 4.2: Daily 5-S Activity for the Workshop — 96
  - Annex 4.3: Daily 5-S Activity for the Office — 97
  - Annex 4.4: 5-S Implementation Plan — 98

5 Business Process Re-engineering — 99
- Introduction: What is BPR? — 99
- Service Industry and Quality — 104
- Best Practice Benchmarking (BPB) — 114
- Total Quality Marketing — 118
- Total Quality Purchasing — 125
- Why is BPR Useful? — 130
- How to Implement BPR — 131
  - Chapter Summary — 134
  - Hands-on Exercise — 134

6 Quality Control Circle and Problem Solving — 135
- Introduction: What is a QCC? — 135
- The Japanese QCC Experience — 136
- The Seven Quality Control Tools — 138
- Problem Solving — 155
- The S-S Problem Solving Method — 156
- The S-S Method and the Seven QC Tools — 165

Why Do We Need QCCs?		167
How to Establish QCCs		169
QCCs and Employee Empowerment		171
Chapter Summary		174
Hands-on Exercise		174

7 ISO 9000 and Quality Audit   175

Introduction: What is ISO 9000?	175
The Content of ISO 9001	177
Quality Audit	181
Why are ISO 9000 and Quality Audit Needed?	186
How to Implement ISO 9000	188
Benefits of Implementing ISO 9000	196
Chapter Summary	200
Self-Assessment Exercise 7.1 : Interpretation of the ISO 9001 Clauses 4.1-4.20	201
Self-Assessment Exercise 7.2 : ISO 9001 Quality Auditing	202
Self-Assessment Exercise 7.3: ISO 9001 Mock Audit	204
Hands-on Exercise	209
Annex 7.1 : List of the ISO 9000 Family	210
Annex 7.2 : The 20 Clauses of ISO 9001:1994	211

8 Total Productive Maintenance   213

Introduction: What is TPM?	213
The Content of TPM	215
Why is TPM useful?	220
How to Implement TPM	223
Chapter Summary	224
Hands-on Exercise	224

9 TQM Kitemarks   225

Introduction: What are TQM Kitemarks?	225
ISO 9004-4: 1993	226
Japan's Deming Prize	226
USA's Malcolm Baldrige National Quality Award (MBNQA)	228
The European Quality Award (EQA)	232
Why are TQM Kitemarks Useful?	239
How to Achieve TQM Kitemarks	241
Chapter Summary	242
Hands-on Exercise	242
Annex 9.1 : ISO 9004-4:1993 Outline (Clause 4-7 only)	243
Annex 9.2 : The European Quality Award Checklist	245

10	**TQMEX in the Real World**	**249**
	Validation of the TQMEX Model	249
	Stage 1 Results: The UK Companies Survey	251
	Stage 2 Results: The UK and Japanese Leading Companies Survey	262
	Significance of ISO 9000 in TQMEX Implementation	270
	How to Implement TQMEX	273
	Chapter Summary	275
	Hands-on Exercise	275

REFERENCES 276

SUGGESTED ANSWERS 281

SOURCES OF INFORMATION AND ADVICE 284

INDEX 289

MINI-CASES

1.1:	Some Examples of Corporate Vision and Mission Statements	8
1.2:	Strategy for Total Quality	9
1.3:	Quality Problems in the Service Industry	15
2.1:	Zytec Implements Deming's 14 Points	28
3.1:	On Commitment	53
3.2:	Effective Communication of Quality	55
3.3:	Quality Improvement Storyboard at Komatsu Company	57
4.1:	My Experience in Throwing Things Away	64
4.2:	Even 5 Minutes Can Make a Difference	89
4.3:	Implementing 5-S at SIRIM	89
4:4:	Implementing 5-S at the Wellex Corp. in the USA	90
5.1:	Service Quality at Scandinavian Air System	107
5.2:	The School of Quality	112
5.3:	Rank Xerox's BPB	117
5.4:	British Telecom Goes into TQMar	119
5.5:	TQMar - Promotion: TV Advertising	123
5.6:	Lucas Industries achieving success through TQMar	124
5.7:	Total Quality in Purchasing	128
5.8:	Boots The Chemists goes for BPR	130
5.9:	BPR at Co-op Bank -- Improving Personal Customer Service	133

6.1:	7-Eleven Using Quality Control Tools as Competitive Advantage	154
6.2:	Johnson Electronic Co. Ltd. of Hong Kong	166
6.3:	A Medium-sized Plastic Injection Moulding Factory	167
6.4:	Hydrapower Dynamics Ltd. of Birmingham	168
6.5:	Employee Empowerment at Lyondell Petrochemical	172
7.1:	Proposals to Prevent Collapse of High-rises	179
7.2:	Benefits of ISO 9000 Implementation	196
7.3:	MBI's ISO 9002 Employee Attitude Survey	197
7.4:	Bowater Containers Southern Goes for ISO 9000	198
7.5:	DTI Import Licensing Branch Goes for ISO 9000	199
7.6:	Praxis Software Engineering Company Goes for ISO 9000	199
8.1:	1986 PM Excellence Award Winner: Toyota Auto Body Co.	218
8.2:	1984 PM Excellence Award Winner: Fujikoshi	219
9.1:	A 1991 MBNQA Winner: Zytec Corporation	231
9.2:	Girobank Wins the British Quality Award in 1991	235
9.3:	Short Brothers Plc Wins the British Quality Award in 1992	237
9.4:	Rover Group Wins the first UK Quality Award	238
9.5:	TNT Express (UK) Ltd. Wins the first UK Quality Award	239
10.1:	TQM Promotion by PHP Institute Inc., Tokyo	269

FOREWORD

The prevailing areas of management concern are quality, cost and productivity. People often talk about quality culture rather than cost and productivity. Quality has been integrated in human history for much longer than cost and productivity, and it is the major concern for both supplier and customer. From these reasons, quality issues cover the soft (people) aspects of business while cost and productivity carry the hard (process) notion in themselves. Therefore, quality improvement will have greater motivating impact on employees than the call for cost reduction and productivity increase.

Quality, with its special features, goes beyond cost and productivity by integrating and directing all value-adding business activities towards a common goal. We also know that when quality is improved in a creative way, cost will be reduced and productivity increased. The opposite is not always true.

To cope with the competition in the open global market, the importance of quality management and customer-focused strategies is increasing. In addition, quality is the common concern of every employee within the company, and it is one of the most powerful motivators for their good work. Continuous improvement *(Kaizen)* is the major driving force in successful businesses. Implementation of companywide improvement plan is based on sound action plans accepted by everyone in the company using quality as the main focus.

Quality is often classified into must-be quality (backward-looking quality) and attractive quality (forward-looking quality). Improving the must-be quality is effective for reducing costs of poor quality. On the other hand, improvements in attractive quality contribute to the increase of profit by expanding the market size and share of the product and service.

The TQMEX model explained in this book is an important proposal: the author explains that this model is a sequence of steps arranged logically to serve as a guideline for implementation of a process in order to achieve the ultimate goal. The range of activities of TQM is broad and well covered. I am particularly pleased to see that 5-S is used as the first step towards TQM, as it has been widely adopted amongst Japanese organisations. The readers should recognise that all the steps in this model are equally important. By moving from one to another, we do not replace the previous activity with more important one, but our quality improvement changes its focus.

In this book many interesting mini-cases, exercises and self-assessment checklists are allotted to each chapter, which help the readers understand the material and apply the knowledge. It is essentially important that we employ Plan-Do-Check-Act cycle in accumulating our problem-solving abilities and knowledge. This is the source of our energy for continuous improvement and significant breakthroughs.

Yoshio Kondo,
Professor Emeritus, Kyoto University, Japan, (1971 Deming Award winner).

FOREWORD

I am greatly honoured to be asked to write this Foreword to Sam Ho's book. I am of course glad to see the relatively long section on Dr W Edwards Deming's work in Chapter 2. Sam Ho's concept of producing an *integrated approach* to TQM is commendable. One of the main problems with common attitudes to TQM is that they are generally too limited. In many organisations there is too strong a concentration on a proliferation of tools and techniques, and far too little on the transformation of culture which is necessary to allow the true potential of those tools and techniques -- and much else -- to be realised. Dr Deming thoughtfully referred to tools and techniques as being "essential but unimportant". You would see very little of them in his celebrated four-day seminars. Even if you saw them, his concentration would be wholly on interpretation and understanding rather than on technical details.

Dr Ho has been kind enough to reproduce a wise statement that he heard me use in a presentation in April 1995: "The control chart is no substitute for the brain." This statement came from a lecture that Dr Deming delivered to top management in Japan in the summer of 1950 - nearly 30 years before the Western world started paying any significant attention to his teaching. I sometimes express the emphasis on the role of tools as being to help the human thought process (rather than attempting to replace it, as seems to be the case in some companies) by referring to them as a brain *pacemaker* rather than a brain *bypass*.

Dr Ho has devoted several pages of Chapter 6 to the use of control charts in the context of problem solving and quality circles. As Dr Deming clearly pointed out in his final book, *The New Economics for Industry, Government, Education*, the potential gains from applying principles of the statistical control of quality (which he simply defined as "knowledge about common causes and special causes") in those areas, large though they may be in absolute terms, still represent only the tip of the iceberg. He estimated that only 3% of the possible rewards could come from such sources. The other 97%, the big gains, await the application of the same theory and practice to the whole organisation, particularly to the *management of people*.

Total Quality Management is a fine phrase. In many minds it is also a fine concept. However, because the term has now become so familiar, there can be a tendency to let the words trip too lightly off the tongue. I recommend the reader to think long and hard about the meaning of each of the three words, as Dr Ho invites you to do early in Chapter 1. Any journey must have a starting point. I am simply advising you that the destination is much further away -- but also that travelling toward it is a much more exciting prospect than many who use the words contemplate.

Henry R. Neave, PhD, FSS,
Director of Education and Research, and
Founder of the British Deming Association.

PREFACE

Many companies work very hard to produce high standard goods and services. Quality is a vital issue in contemporary business and is becoming a distinctive competitive advantage. Everyone talks quality today. The theory and experience are brought together and form an abundant amount of knowledge. Each company that has chosen the path of total quality towards excellence has its own approach. The fame of the Japanese as a quality nation is prevailing. Competition transcends national boundaries. Global market, benchmarking, ISO 9000 are points of debate in business environment at the end of this century.

There are so many areas to tackle. Companies ask: how to start and where does this all lead to? This book is the result of the effort to put the best quality practices together in a sequential model which can be used as a step-by-step guide for the companies who want to achieve TQM. It is argued that organisations are different and there is no universal approach to total quality. This book attempts to offer something to everyone.

It has been written to meet the needs of those who are dedicated to quality improvement and those who want to learn more about TQM: from the chief executives to the front line workers of any organisation, be it large or small, private or public, manufacturing or service. It has been designed as a manual for day-to-day quality routines but can also serve as a textbook for university students on management and professional courses.

It is difficult to specify when the idea for this book was born. Perhaps as far back as 1986 when sponsored by the Asian Productivity Organisation, I spent a year doing research on technological development of the five tigers of Asia (Japan, Hong Kong, Korea, Singapore and Taiwan). There I realised the importance of the notion of continuous improvement. In order to emphasise this, I have written the Japanese word (same as Chinese) for Continuous Improvement, *Kaizen,* on the front cover of this book.

The 5-S technique has for surprisingly long time been a secret to the Western World. It originated in Japan, where clealiness and orderliness of the working place are companies' main concerns. It triggered my interest while I was doing research in Japan in 1987. Since then, I have been expanding the theoretical and practical aspects of this technique. Being convinced that it works as a powerful quality tool, I have emphasised its importance in this book.

The TQMEX Model originated in my working as the first Quality Expert to the Malaysian Government via the Standards and Industrial Research Institute of Malaysia (SIRIM) in 1993. The role was to develop a 5-year Quality Plan for the improvement of the ISO 9000, 5-S and TQM services offered by SIRIM to enterprises in Malaysia. At SIRIM, I developed a model of excellence named "TQMEX" which has been accepted by the Director General, installed, promoted and implemented in Malaysia.

The content of this book follows the flow of the TQMEX Model. The first chapter is devoted to the fundamentals of TQM. Chapter 2 summarises the important teachings of the quality gurus. Chapter 3 explains how TQMEX was developed and each element of the TQMEX is discussed in depth in Chapters 4--9. The last chapter validates TQMEX based on the surveys conducted in the UK and Japan, and finishes with an implementation plan. The book contains a number of diagrams and charts as well as 40 mini-case studies related to the TQMEX Model elements.

Purely for reasons of convenience, 'he' and other forms of the masculine gender are used throughout. There just is no satisfactory neutral form and 'he or she' etc. always seems clumsy. There is, of course, absolutely no reason why women should not be the prime movers in instigating and managing TQM.

I am indebted to Svetlana Cicmil for her constructive criticism, time and effort spent on advising on the structure and content of this book. I would like to thank Yoshio Kondo and Henry Neave for their valuable comments. I am thankful to Christopher Fung who has helped me to conduct the questionnaire survey based on the TQMEX. I am also grateful to my wife Catherine, my daughter Olivia and my recently deceased mum who always understand and support me. Finally, I would like to thank all the authors quoted, whose research and ideas have enriched this book.

Samuel K. Ho,
Sep 1995

ABBREVIATIONS

5-S	Five Japanese words: *Seiri, Seiton, Seiso, Seiketsu and Shitsuke*
ANOVA	ANalysis Of VAriance
APO	Asian Productivity Organization
BPR	Business Process Re-engineering
BSI	British Standards Institution
CEO	Chief Executive Officer
CWQC	Company Wide Quality Control
EFQM	European Foundation for Quality Management
EQA	European Quality Award
ISO	International Organization for Standardization
IT	Information Technology
JIT	Just-In-Time (Production System)
JIPE	Japan Institute of Plant Engineers
JUSE	Japan Union of Scientists and Engineers
Kaizen	The Japanese word for 'Continuous Improvement'
LCL	Lower Control Limit
MBNQA	Malcolm Baldrige National Quality Award
PDCA	Plan-Do-Check-Act Cycle (Deming Cycle)
PDSA	Plan-Do-Study-Act Cycle (Revised Deming Cycle)
QAT	Quality Action Team
QC	Quality Control
QCC	Quality Control Circle
QI	Quality Improvement
QMS	Quality Management System
QSC	Quality Steering Committee
Servqual	Service Quality Model (by Parasuraman A.)
TPM	Total Productive Maintenance
TQ	Total Quality
TQM	Total Quality Management
TQMar	Total Quality Marketing
TQMEX	The TQM EXcellence Model used in this book
TQPur	Total Quality Purchasing
UCL	Upper Control Limit
ZD	Zero Defects

Chapter 1

Introduction to TQM

Topics:

- ▶ The Japanese Quality Miracle
- ▶ What is quality and TQM?
- ▶ TQM and corporate strategy
- ▶ The four pillars of TQM
- ▶ Some important TQM concepts
- ▶ Why is there a need for TQM?

The Japanese Quality Miracle

Much has been said about the Japanese economic miracles over the past decade. This trend continues despite the recent appreciation of Japanese Yen. The western world might have thought that this would make the Japanese products less competitive. Surprisingly, this has not been the case. There is still no sign of their enormous trade surplus coming down!

David Ricardo is recognised for his economic theory on comparative advantage which predicts that economic activities will move to places where overall natural resources are abundant and market is near. Yet this cannot explain the fact that Japan exports to the USA ten times more than Brazil although the latter has the same population and is about ten times closer to the USA. To explain this aspect, we have to consider the notion of "Dynamic" Comparative Advantage. This implies that comparative advantage is not a gift from nature alone, but a dynamic entity designed by people and includes investment in education, public facilities, technology, export promotion, etc.

Total Quality Management (TQM), an intangible investment not regarded as such by the economists is equally, if not more, important for economic growth. Japan could not be as successful without its huge investment in TQM. During 1987 I had an opportunity to study the Japanese economy success under the Oshikawa Fellowship Scheme sponsored by the Asian Productivity Organisation (APO). The scheme included research visits to representative organisations in the manufacturing and services sectors in Japan. Useful information was also obtained from the Ministry of International Trade and Industry (MITI), which has earned its name as the *Invisible hand of the Japanese economic miracles*. It was observed that terms like 5-S, QCC, TPM and TQM were part of the daily language and activities in most of the firms visited. Even before one walked inside the company, the environment would project an impressive image of quality. Inside the factories and offices places were neat and tidy with a lot of slogans, charts and pictures developed by the workers themselves. Even in their own work area, the workers tended to display some sort of control charts and Pareto diagrams (explained in Chapter 6) to ensure that their processes were under control and to show that they were constantly improving them. *TQM was integrated into companies' management practices and operations.*

Rooted in the research and teachings of American quality gurus and the home grown management synthesis of Japanese quality gurus (see Chapter 2), the TQM message empowered Japanese managers and gave them the quality vehicle to establish global trade supremacy. The quality gurus told Japanese managers and politicians that the only way they could survive as an island nation with virtually no natural resources was to make quality products and service the national imperative. Most of the older generation of Japanese still have the picture in their mind that there is a line of super-tankers joining hull to stern all the way from the Persian Gulf to Tokyo Bay!

And this they did, producing steel, ships, motorcycles, cars, trucks, drugs, medical equipment, transistor radios, stereo components, television sets, video recorders, cameras, camcorders, home entertainment centre, calculators, computers, office equipment, musical instruments, hand tools, machine tools, tyres, electric motors, food processors, microwave ovens, sports equipment, heavy machinery, computer games, industrial robots, electron microscopes and micro-chips. They measure their defects in 'parts per million' rather than the traditional Western percentages. In all these sectors the Japanese sell their products on quality as well as price. They perfected the idea of continuous improvement. How did they do all these?

History of Quality Control in Japan

The pre-war reputation of Japanese products was far from being favourable. "Made in Japan" label was synonymous for cheap and shoddy work. It was only in the post-war era that modern quality control was introduced in the sense of statistical quality control. In 1949, the Union of Japanese Scientists and Engineers (JUSE) established a Quality Control Research Group, thus paving the way for fully-fledged quality control education, dissemination, and practice in Japan. They realised that quality products cannot be produced without a commitment from office and factory workers, salesmen, and supervisors alike. The Group promptly set about providing quality control training for these front-line people.

At the same time, JUSE invited such prominent Americans as Dr. W. Edwards Deming and Dr. J.M. Juran to come to Japan to lecture on quality control methodology. Although these American teachings were valuable, they were not accepted at face value. Rather they were adapted to suit the distinctive Japanese social background and other circumstances, and an effort was directed towards making the entire labour force quality conscious.

Chapter 1: Introduction to TQM

In 1956, broadcast media were also mobilised for this educational effort, first radio and later television. Then in 1960, *Quality Control Text Book for Foreman* was published, followed by the *Quality Control for the Foreman Monthly Magazine* in 1962. In the same year, the first quality control circle was born. In 1983, with a circulation of 150,000 the magazine has become a major force for cross-education.

During the period 1960-1980, quality control evolved from statistical quality control to Company Wide Quality Control (CWQC) which was characterised by the following three points:

- All department participation (including subcontractors, sales companies, and maintenance/service companies).
- All employee participation (from top management to line workers and salesmen).
- Integrated process control (including quality, profits, delivery, safety, and contribution to society).

CWQC is in effect the Japanese equivalent for TQM. CWQC in Japan began with the manufacturing industries, but it has spread since the mid 1970's on non-manufacturing industries such as construction, banking, insurance, retail sales, hotels and restaurants, transportation, and public utilities.

What is the Lesson for the Rest of the World?

"Japanese and American management is 95% the same and differs in all important respects."
-- *Takeo Fujisawa, Cofounder of Honda Motor Co.*

Although native to the USA, the ideas of Deming and Juran were ignored in America. In the 1950s and 60s American businesses could sell everything they made to a world hungry for manufactured goods. The emphasis of American and most Western manufacturing industries was on maximising output and profit. In their existing markets quality of the goods had a low priority. In the early 1970s, when the oil crisis came, competition became intensified. The American companies started to take the quality message more seriously when they discovered that they had lost both markets and market share to the Japanese. They wondered why it was that consumers preferred Japanese products.

During the 1980's, the quality movement came back to America. The turning point happened in 1980 when the National Broadcasting Corporation produced a documentary called "If Japan can, why can't we?". This programme highlighted the dominance of Japanese industry in many US markets. The latter part of the programme featured Deming and his contribution to Japanese economic success. It sent a shock-wave through the country, and to other parts of the world. People began to realise the importance of Deming's work in Japan and the importance of quality in modern day businesses. The shock-wave entered Western Europe and spread across the Far Eastern countries very quickly.

The benefits of TQM include greater competitive advantage and massive financial savings. *Quality cost analysis* provides an organisation with valuable information for cost-saving opportunities and encompasses the costs of prevention, appraisal, internal failure, external failure, exceeding customer requirements and lost opportunities.

Although the manufacturing industry was the first one to take advantage of the new movement, TQM proved effective in the service industry, as well. In fact, whenever an organisation has a sequence of activities directed towards a defined end result, its business processes can be analysed and improved by TQM techniques.

As promoted by the various government bodies and professional institutions, the rest of the world is making evolutionary changes toward TQM in order to capture the benefits. Getting it right first time, zero defects, prevention, the internal customers, competitive benchmarking, cost of quality, team work, self-management, and self-inspection have become key words and phrases in the move towards TQM. Behind the words and concepts are techniques and actions. Instead of focusing on products and services in isolation, they are analysed in their integral connection with business processes and the people who do the work. New systems and processes are created. The rewards are great for those companies who meet customer requirements because they win repeated business.

What is TQM ?

TQM provides the overall concept that fosters continuous improvement in an organisation. The TQM philosophy stresses a systematic, integrated, consistent, organisation-wide perspective involving everyone and everything. It focuses primarily on total satisfaction for both the internal and external customers, within a management environment that seeks continuous improvement of all systems and processes.

TQM emphasises use of all people, usually in multifunctional teams, to bring about improvement from within the organisation. It stresses optimal life cycle costs and uses measurement within a disciplined methodology in achieving improvements. The key aspects of TQM are the prevention of defects and emphasis on quality in design.

TQM is a necessity. It is a journey. It will never end. It makes Japanese industry a miracle. It is the way to survive and succeed. What does it entail, then? TQM is the totally integrated effort for gaining competitive advantage by continuously improving every facet of an organisation's activities. If we look at the meaning of each word, **TQM can be defined as:**

Total - Everyone associated with the company is involved in continuous improvement (including its customers and suppliers if feasible),
Quality - Customers' expressed and implied requirements are met fully,
Management - Executives are fully committed.

NOTE: *Ideally, everyone in the organisation should be committed. However, according to Deming's research, [1986] some 94% of the problems in quality are caused by management and the system they create. Therefore, commitment by management should come before that of the front-line workers. Totality of managing quality implies that everyone, including the front-line workers, should be involved in the process. Thus the above definition of TQM is a good balance between the ideal and the real world. Various definitions of quality will be discussed in the next section.*

What is Quality?

In the TQM definition above quality is interpreted as "Customer's expressed and implied requirements are met fully". This is a core statement from which some eminent definitions of quality have been derived. They include: *"the totality of features and characteristics of a product or service that bears on its ability to meet a stated or implied need"* [ISO, 1994], *"fitness for use"* [Juran, 1988], and *"conformance to requirement"* [Crosby, 1979]. It is important to note that satisfying the customers' needs and expectations is the main factor in all these definitions. Therefore it is an imperative for a company to identify such needs early in the product/service development cycle. The ability to define accurately the needs related to design, performance, price, safety, delivery, and other business activities and processes will place a firm ahead of its competitors in the market.

In 1992 Crosby broadened his definition for quality adding an integrated notion to it: *"Quality meaning getting everyone to do what they have agreed to do and to do it right the first time is the skeletal structure of an organisation, finance is the nourishment, and relationships are the soul."*

Some Japanese companies find that "conformance to a standard" too narrowly reflects the actual meaning of quality and consequently have started to use a newer definition of quality as *"providing extraordinary customer satisfaction"*. There is a trend in modern day competition among Japanese companies to give you rather more in order to 'delight' you. So when you buy a lamp bulb which has a 'mean time between failure' of 1,000 hours, the Japanese manufacturer will try their best to ensure that you can get at least 20% more. Likewise, when you buy a Japanese brand video tape specifying 180 minutes, it can normally record up to 190 minutes. When you buy a 'mink' coat from a department store in Japan, they would invite you to store the fur coat in their temperature-control room during the hot summer season free-of-charge. They call these extra little things as 'extra-ordinary customer satisfaction' or 'delighting the customers' (see Figure 1.1).

Figure 1.1 Delighting your customer [Lip, 1989]

After you have a good understanding of what quality means, it will be useful to do the Quality Awareness Checklist in the Self-Assessment Exercise 1.1 at the end of this Chapter. Then check your results at the end of the book.

Relationship between TQM and Corporate Strategy

Corporate strategy consists of three key phases. The first phase is the determination of a corporate mission statement which sets the common value for everyone in the organisation. This *mission statement or vision* of the firm should sustain the challenge of time and should remain unchanged for a decade or more. The second phase is defining the strategic options and choosing the optimum one. Normally, this will become the three to five year plan for the organisation. The third phase is the strategic implementation which is also known as operations management. It also defines the short term (three months to one year) plan for the organisation. The question is: Where does TQM fit into Corporate Strategy?

Chapter 1: Introduction to TQM

There are different answers to this question. Most people think that TQM should be linked to operations management. This is by no means a coincidence because before the Japanese developed TQM, they started with industrial engineering, quality control, company wide quality control, and value engineering. Unfortunately, it is not always obvious that during the evolution of TQM, the concept of quality awareness has filtered up the organisational hierarchy. Today, the Japanese managers and directors are so concerned about quality that it has long become their mission. Consequently, in their strategic formulation process, they have used quality as their key mission statement and strategic option. When it comes to strategic implementation, quality has become a routine. The relationship between TQM and corporate strategy is illustrated in Figure 1.2. This approach adds totality to quality, as it is communicated throughout the organisation and spanned over its long term plan.

Figure 1.2 The relationship between corporate strategy and TQM

It is useful at this point to study some of the best examples of corporate mission statements to understand how these companies have built TQM into their declarations (see Mini-Case 1.1).

Mini-Case 1.1:

Some Examples of Corporate Vision and Mission Statements

Example 1: **Dow Chemicals** *(by H.H. Dow, Founder)*
Vision:
"A premier global company, dedicated to growth, driven by quality performance and innovation, committed to maximising our customers' successes, and always living our Core Values."
Mission Statement (Core Values):
- Long-term profit growth is essential to ensure the prosperity and well-being of Dow employees, stockholders, and customers. How we achieve this objective is as important as the objective itself. Fundamental to our success are the core values we believe in and practise.
- Employees are the source of Dow's success. We treat them with respect, promote teamwork, and encourage personal freedom and growth. Excellence in performance is sought and rewarded.
- Customers will receive our strongest possible commitment to meet their needs with high quality products and superior service.
- Our Products are based on continuing excellence and innovation in chemist-related sciences and technology.
- Our Conduct demonstrates a deep concern for ethics, citizenship, safety, health and the environment.

Example 2: **Sea-Land** *(by A.J. Mandl, Chairman and CEO)*
Vision:
"A unique transportation-distribution capability"
Mission Statement:
- Sea-Land is driven by the requirements of its customers.
- Sea-Land's strategy is to strengthen its competitive advantages by linking all global transportation elements into a cohesive system supported by an array of information services that customers clearly value.
- The company's assets and capital, operational enhancements, organisational structure and marketing initiatives are all focused on this goal.
- We are convinced that this strategy secures a vigorous and prosperous future for Sea-Land and each one of its employees.

Example 3: **Matsushita Electric** *(by K. Matsushita, Founder)*
Vision:
"Profits are linked to growth and that investments which promote growth will eventually pay off in the long term."
Mission Statement:
- National service through industry,

Chapter 1: Introduction to TQM   9

- *Fairness,*
- *Harmony and co-operation,*
- *Strive for betterment,*
- *Courtesy and humility,*
- *Assimilation to the society, and*
- *Gratitude.*

Example 4: Leicester Business School, De Montfort University *(by J. Coyne, Head of School)*
Vision:
"By the year 2000 we want to be recognised amongst the leading Business School in Britain. With our critical mass and full range, full function approach to Business we aim to be 'The Best of the Big.'"
Mission Statement:
- *To be recognised as a major provider of high quality teaching, research, management development and consultancy, servicing clients regionally, nationally and internationally.*
- *To provide a supportive working environment, with staff and students enjoying facilities of the highest standards attainable within the resources available.*
- *To contribute fully to the creation of a major international University of the Year 2000.*

■■■

Mini-Case 1.2 is a case study of a British Quality Award firm (see Chapter 9 for details). It is not difficult to find out from the discussion that the firm has implemented all of the four pillars of TQM with good success.

■■■

Mini-Case 1.2:

Strategy for Total Quality   [Ball, 1992]

In November 1991 John Laing plc became the first building and construction company to win the British Quality Award 'for outstanding achievement in implementing a total quality process in the building and construction industry to the satisfaction of customers and with the involvement of employees and subcontractors.' John Laing Group is an international organisation employing over 10,000 people in the UK; in 1990 its turnover was over £1.5 billion. It is one of the top construction, housing, mechanical and civil engineering groups in the UK offering a range of traditional and specialist contracting services for large and small scale contracts.

The Laing approach to quality is based on achieving customer satisfaction, control of quality and continual improvement. Their quality management system has been founded on the commitment of senior management and improved quality awareness and training programmes for employees at all levels in the organisation; co-operation with suppliers also formed a vital part of the total quality approach. This has created a Company culture of business efficiency combined with employee benefits and concern for the wider environment, which has been widely recognised.

The success of the quality improvement of Laing is a result of its quality policy aimed to:

- *Achieve customer satisfaction by completing contracts in the most effective manner and by providing products and services that meet the specified requirements and are in accordance with the need and expectations of customers.*
- *Develop and maintain an economical, practical and documented Quality Management System for each of our Divisions and operating businesses.*
- *Maintain and improve the quality, safety and performance of our products and services, leading to enhanced reputation before clients and the public.*
- *Ensure we manage quality as effectively as we manage time and cost.*
- *Organise and arrange our affairs in such a way that the factors affecting the quality of contract, product and services provided by the company are always under control.*
- *Develop and implement a Group-wide quality improvement process directed at creating committed customers, improving productivity, reducing costs and increasing employee participation in this process.*
- *Use our high standards in Quality Management to increase profitability and market share.*

As an important step towards TQM, Laing registered itself under the ISO 9000 quality management system in the late 1980's. This demonstrates their commitment towards quality and satisfying the needs of their customers. The evidence of customer satisfaction are measured through a variety of methods such as the number of repeated business, the use of postal customer questionnaires, direct interviewing, and third-party surveys. This includes internal as well as external customers. The idea of building models of their new design homes in the warehouse and asking previous purchasers to come along and critically assess these models has been extremely beneficial.

Laing cannot survive without satisfying their customers. The results of which are the number of awards they are consistently gaining -- such as Manager of the Year awards, Apprentice of the Year awards, Construction Industry Supreme Awards, Contractor of the Year awards, Housebuilder of the Year awards.

■■

Some Important TQM Concepts

The Quality Chain

As shown in Figure 1.5, suppliers and customers not only exist outside an organisation, but inside too (i.e. internal customers). There is a series of supplier/customer relationships. These relationships serve as important interfaces in the quality chain. Failure to meet the requirements in one part of a quality chain will affect the other. For example, if the sales department has promised a certain delivery date, and the production department fails to comply, then both the internal customer and external customer will be dissatisfied. The cause of the nonconformance could be production, marketing or both due to poor communication. To avoid this, TQM needs to be practised throughout the organisation.

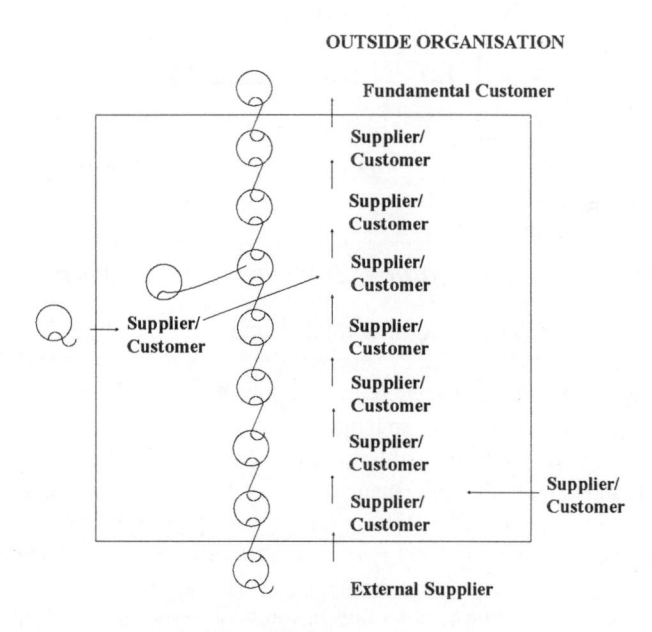

Figure 1.5 The Quality Chain

The Eight Stages of the Industrial Cycle

Figure 1.6 gives a clearer idea of the supplier/customer relationships. It shows the stages a product needs to go through before reaching a consumer's hands. To make sure that the right products are produced at the first time with zero defect, co-ordination between all the stages is important. As a result, TQM is the means to assure that customers will get what they want right first time, each time and every time.

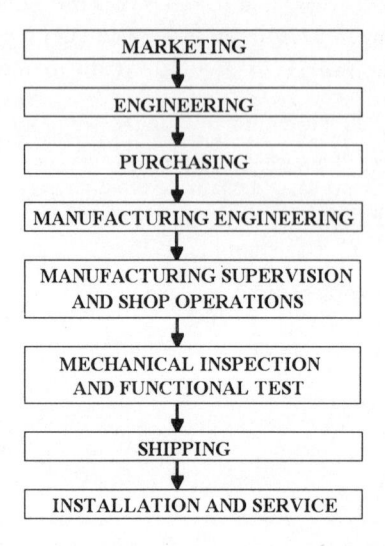

```
MARKETING
   ↓
ENGINEERING
   ↓
PURCHASING
   ↓
MANUFACTURING ENGINEERING
   ↓
MANUFACTURING SUPERVISION
AND SHOP OPERATIONS
   ↓
MECHANICAL INSPECTION
AND FUNCTIONAL TEST
   ↓
SHIPPING
   ↓
INSTALLATION AND SERVICE
```

Figure 1.6 The eight stages of the industrial cycle

The ISO 9004-4:1993 Quality Improvement Guidelines

Continuous Improvement is an important theme of TQM. It is also the objective of the new ISO 9004-4 Standard. The background of the ISO 9004-4 Standard for Quality Improvement is discussed in Chapter 9. The Standard is a guidance document produced to help management to organise and promote quality improvement. It starts off by endorsing the fact that without total dedication from the top management to the principles of TQM, the chances of effective applications of the concepts throughout the organisation are weak, and the necessary change to management and work attitudes would be difficult to achieve.

The main principles of the Standard recognise that customer satisfaction, health and safety, the environment and business objectives are mutually dependent and that each business can be broken down into a series of processes. Above all, TQM involves investment in both time and people. The rewards can be substantial, but the commitment must be total.

The Standard deals with the implementation of a continuous quality improvement process which can be applied to every aspect of the organisation. It concludes with an informative list of the most common tools and techniques used. A list of tools and techniques suggested by the ISO 9004-4 are:-
- Benchmarking
- Process Flowchart (incorporating Tree diagram)

Chapter 1: Introduction to TQM 13

- Data Collection Form
- Histogram
- Pareto Analysis
- Cause and Effect Diagram (incorporating Brainstorming)
- Scatter Diagram
- Control Chart

(Benchmarking will be discussed in Chapter 5, and all other tools will be discussed in Chapter 6 under the 'Seven QC Tools'.)

After you have a good understanding of what TQM means, it will be useful to do the TQM Awareness Checklist in the Self-Assessment Exercise 1.2 at the end of this Chapter. Then check your results at the end of the book.

Why is there a need for TQM?

"In order to compete in a global economy, our products, systems and services must be of a higher quality than our competition. Increasing Total Quality is our number one priority here at Hewlett-Packard."

-- *John Young, President of Hewlett-Packard*

From the four examples of corporate visions and mission statements in Mini-Case 1.1, it is apparent that all these companies want to provide good quality goods and services to their customers. The end result is that they will enjoy prosperity and long-term growth. This is why Deming considered the contents of a diagram presented in Figure 1.4 the most important objective of quality management. In fact the diagram is no longer new to the Japanese. Since the 1950's onward the management in many Japanese firms have adopted this chain reaction. Management and workers have the same aim -- quality. This chain reaction offers one benefit after another on quality improvement. It is difficult to accept initially that improving quality can improve productivity. However, Japan's marvellous economic achievement is undeniable solid evidence. So TQM is essential in activating Deming's chain reaction. (More of Deming's ideas can be found in Chapter 2.)

Important aims of TQM include the elimination of losses and reduction of variability. For example, Toyota Inc. in Japan organises their athletic meeting at their own stadium annually. Every year, they have to hire a contractor to re-paint the lines on the track and field. In one year, the Truck Painting Shop staff volunteered to do it themselves. As a result, they only have to spend 5% of the contracted cost to buy the painting materials. The job was finished on time, with good quality, and everyone took pride in it. In their common staff canteen, they also standardise one kind of container -- the same bowl is used for soup, tea, rice, dessert, etc. As a result, there was a significant saving. More important is that each staff member has been reminded of the importance of standardisation and elimination of losses every time they go into the canteen. Furthermore, TQM advocates the development of employee, supplier and customer relationships, a philosophy very much emphasised by Toyota in their Just-In-Time Production System which will be discussed in Chapter 5. Finally, the philosophy of TQM is based on an intense desire to achieve victory.

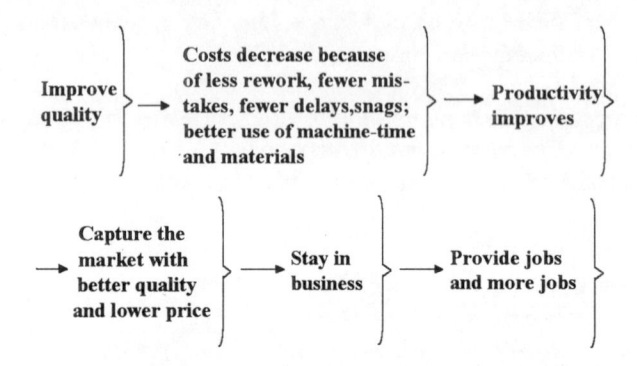

Improve quality → Costs decrease because of less rework, fewer mistakes, fewer delays, snags; better use of machine-time and materials → Productivity improves

→ Capture the market with better quality and lower price → Stay in business → Provide jobs and more jobs

Figure 1.4 Deming's Chain Reaction

Quality Problems in Industry

Quality problems can be found in all sectors of industry: manufacturing, service, public service, education and training, etc.

Employing more inspectors, tightening up standards, developing correction, repair and rework teams will not promote quality. Nevertheless, these activities still exist in today's industry. There are some typical problems that people deal with most of the time. They include:
- Misspelled written material
- Wrong data entry and calculation
- Time lost due to equipment failures
- Change work instruction due to error in design
- Billing errors
- Missing files
- Contract errors
- Stock shortages
- Hotel guests taken to unmade rooms
- Reservations not honoured
- Purchase order changes due to error
- Rejections due to incomplete description
- Checking things because of lack of trust
- Rectifying and reworking
- Apologising to customers
- Scrap, returns

Chapter 1: Introduction to TQM

- Warranty claims, service
- Late reports

After you have a good understanding of the potential quality problems in your workplace, it will be useful to do the F-test in the Self-Assessment Exercise 1.3 at the end of this Chapter. Then check your results at the end of the book.

One interesting rule-of-thumb in quality is called the **1-10-100 Rule.** If someone produces defective work, and rectifies it immediately, it only costs him another equal effort to do so. If it has slipped to his internal customers and he wants to rectify it then, it will cost 10 times more effort. If, unfortunately, it has passed on to the external customer, then he has to pay around 100-fold in order to get the error rectified and the adverse consequences that follow. This is particularly significant for the service industries as most of their work directly deals with external customers. They have less chance to rework internally. So quality is compulsory for survival (see Mini-Case 1.3).

There are also some interesting findings by Crosby [1992] and Hakes [1991]. Crosby finds out that instead of learning to understand the process and eliminate the causes of errors, some companies concentrate on devising bigger and better testing to find the bad components and keep the good ones. He also estimates that manufacturing companies waste 25% of turnover on doing things wrong or reworking.

Hakes mentions several quality problems too. He points out that companies used to detect defects and errors in their products and services and then congratulated themselves on taking remedial action to put them right. They continue fire-fighting and rectifying the same problems desperately week after week, month after month, and year after year! He has found out that all the quality gurus believe that over 80% of quality problems are caused by management and fewer than 20% are caused by workers.

■■

Mini-Case 1.3:

Quality Problems in the Service Industry

Imagine what would happen if an emergency service neglected to attend to issues of quality. The following scene illustrates how professionals in the public sector make quality a priority:

The warning bell in the District Fire Hall sounds as John shakes himself awake. It is 3:00 a.m. and a two-alarm fire is burning out of control in a retired people's home a mile away. The on-duty fire crew slowly pull themselves from their warm cots and fumble for their coats and boots. "Where are my gloves?" wonders John as he saunters toward Engine #1. Recently he noticed that the old fire-engine had been giving signals of battery failure. "Should report that," thought John idly.

John gives a final try to secure a water hose on the truck. A loose nozzle falls from the attached hose and clatters to the floor. "Should fix that," he mutters, as he climbs into the cab and yells for everyone to climb aboard. The crew grumbles as they climb on the tail board. Joe, the rookie, is nowhere to be seen.

"Hey, Joe, let's go!" yells John. Joe appears, looking sleepy and confused.

"I thought this was another false alarm," he says.

"Get on, Joe, this is the real thing," yells John, slightly annoyed. John turns the key to start the engine and it grinds hesitantly, then dies. "Should get this fixed," Patrick growls. "Okay, everyone, battery trouble. Unload and reload on Engine #2."

Engine #2 swings slowly from the firehouse. Joe hangs precariously from the safety bar scratching his head. "I sure thought this was a false alarm."

Analyse the Example:
If this story had been true, what were the crew's chances of getting to the fire before the building was destroyed? In your opinion, what quality problems did the District Fire Brigade have? Check your responses from the list below.

- ☐ No clear guidelines for quality standards
- ☐ Poor maintenance
- ☐ Poor team spirit
- ☐ Inadequate training
- ☐ No sense of urgency
- ☐ Lack of communication
- ☐ Unconcerned leadership
- ☐ No preventive thinking

Add your own:
- ☐ _____
- ☐ _____
- ☐ _____

Chapter 1: Introduction to TQM

Chapter Summary:

▶ *The Japanese Quality Miracle has happened because they integrated TQM into their management and operations.*
▶ *TQM is the commitment and action from everyone in the organisation to satisfy their customers, both internal and external.*
▶ *TQM and corporate strategy are closely linked at all levels -- corporate mission, strategic formulation and strategic implementation.*
▶ *Important TQM concepts are based on the supplier-customer chain, eight stages of industrial cycle, and the quality improvement guidelines of the ISO 9004-4.*
▶ *TQM is needed to support Deming's Chain Reaction. Quality problems in industry prevail and the situation will not change unless TQM is in place.*

Self-Assessment 1.1: Quality Consciousness Checklist

Quality begins with an awareness of some basic quality concepts. For example, as a consumer, you probably have developed a 'quality consciousness' over the years. Remember how you liked the mint-green toothpaste better than the white colour, because it gave you an additional mint smell after brushing your teeth. Later you made many life choices based on quality: where you lived and worked, who your friends were, what lifestyle you wanted. The following exercise is to help you to identify your quality consciousness. Please consider each of the following statements and mark it **True (T) or False (F)** based on your current awareness of quality at work and in your personal life.

_____ 1. Quality is preventing problems rather than picking up the pieces afterward.

_____ 2. Quality can always be improved.

_____ 3. The **KISS** (Keep It Short and Simple) method is the best way to ensure quality.

_____ 4. The most important reason for a quality programme at work is to have satisfied customers.

_____ 5. Constant attention to quality is unnecessary.

_____ 6. First impressions aren't important in creating a quality environment.

_____ 7. Quality is the little things as well as the big things.

_____ 8. A quality programme must have management support to be successful.

_____ 9. Quality guidelines are best communicated by word-of-mouth.

_____ 10. Most people want to do quality work.

_____ 11. Customers pay little attention to quality.

_____ 12. A quality programme must integrate with the organisation's goals and profit plans.

_____ 13. Quality means conformance to standards.

_____ 14. Quality should operate in all parts of a business.

_____ 15. Personal quality standards and business quality standards have little in common.

_____ 16. Quality requires commitment.

_____ 17. Quality relates to the process as much as to the goal.

_____ 18. People who talk about quality are idealists.

Self-Assessment 1.2: TQM Consciousness Checklist

Just like Quality, TQM begins with an awareness of some basic TQM concepts. The following exercise is to help you to identify your TQM consciousness. Please consider each of the following statements and mark it **True (T) or False (F)** based on your current awareness of TQM at work.

_____ 1. A philosophy
_____ 2. A quick fix
_____ 3. Goodness
_____ 4. Conformance to perfection standards (such as 'zero defect')
_____ 5. Prevention
_____ 6. Merely inspection
_____ 7. Following specific guidelines
_____ 8. A 'close enough' attitude
_____ 9. A motivational programme
_____ 10. A lifelong process
_____ 11. Commitment
_____ 12. Coincidence
_____ 13. Supported by upper management
_____ 14. A positive attitude
_____ 15. A watchdog mentality
_____ 16. Agreement
_____ 17. Randomly adopted
_____ 18. Do your own thing
_____ 19. Willing communication

Chapter 1: Introduction to TQM

____ 20. Understanding your processes
____ 21. Isolated data
____ 22. Guessing
____ 23. Detecting errors in end products
____ 24. Do it right first time

Self-Assessment 1.3: The F-Test

(Adapted from Henry R. Neave's book: *The Deming Dimension*, page 300, SPC Press, Knoxville, Tennessee, 1990.)

Suppose that the letter F represents a defective item, and all other letters good items. Don't look right now, but if you turn the page you will find displayed a straightforward 17-word sentence; it begins with the word: "FINISHED ... " -- there you are, one defective item straightaway! Give yourself 15 seconds to read through the sentence and count the number of F's, *once only*. Easy enough? OK, check your watch, turn over the page, and try it!

Hands-on Exercise:

1.1 Try to find out the exact requirements of your customers (you may have different requirements from different customers).

1.2 Develop a quality policy (mission statement) for your own organisation.

1.3 Develop your own personal quality policy (including a statement of how you can satisfy your internal customers).

> FINISHED FILES ARE THE RESULT OF YEARS
> OF SCIENTIFIC STUDY COMBINED WITH
> THE EXPERIENCE OF MANY YEARS.

Chapter 2
The TQM Gurus' Ideas

Topics:

- Who are the TQM Gurus?
- W. Edwards Deming *(management philosophy and systems)*
- Joseph M. Juran *(quality trilogy)*
- Philip Crosby *(zero defects and cost of quality)*
- Kaoru Ishikawa *(simple tools, QCC, company-wide quality)*
- Shigeo Shingo *(Fool-proofing)*
- Yoshio Kondo *(four steps for making creative and quality work)*
- Why are TQM Gurus' ideas useful?
- How to implement TQM Gurus' ideas

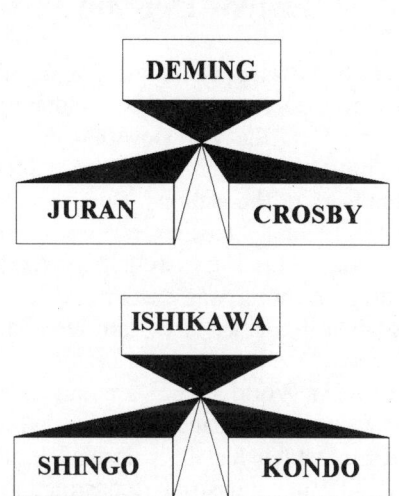

Introduction: Who are the TQM Gurus?

If quality is important, so are the people that propound it. A guru, according to the Oxford Dictionary, is "a respected and influential teacher or authority". A Quality Guru certainly should have these traits in addition to being *a charismatic individual whose concept and approach to quality within business and life, has made a major and lasting impact.* Since TQM originates from Japan, it is natural to identify gurus who have made an impact on the Japanese quality miracle (see Chapter 1). Two groups of Quality Gurus with their key ideas can be identified as follows:

1. The Americans who brought the messages of quality to Japan and the world
   -- W. Edwards Deming (Deming cycle, the 14 points of management and the system of profound knowledge)
   -- Joseph M. Juran (quality trilogy: planning, control and improvement)
   -- Philip Crosby ('do it right first time' and quality cost)
2. The Japanese who developed new concepts in response to the Americans' messages
   -- Kaoru Ishikawa (quality control circles/tools, and company-wide quality)
   -- Shigeo Shingo (foolproofing and zero defects)
   -- Yoshio Kondo (four steps for creative and quality work)

Brief biographies of these gurus and their main ideas will be discussed in greater detail in the following sections.

The American Quality Gurus' Ideas

W. Edwards Deming

Dr. W. Edwards Deming is possibly the most famous quality guru. Born in 1900, Deming was awarded his doctorate in mathematical physics in 1928. He then worked in the US Government Service, particularly dealing with statistical sampling techniques. He applied Shewhart's statistical process control concepts to his work at the National Bureau of the Census. Routine clerical operations were brought into statistical process control in preparation for the 1940 population Census. This led to six-fold productivity improvements in some processes. As a result, Deming was encouraged by some people in the US Government to run statistical courses to explain his and Shewhart's methods in the US and Canada [Neave 1995].

After World War II, Deming was sent to Japan by General MacArthur as an adviser to the Japanese Census. Whilst there, he met some of the members of the JUSE, which had been formed in 1946 by Mr. K. Koyanagi with the task of helping the industrial reconstruction of Japan after the devastation of the war. Mr. Koyanagi invited Deming to come back to Japan to teach on quality control methods for industry. Deming returned in June 1950 to talk to the presidents of 21 leading companies representing some 80% of Japan's capital investment.

The Japanese listened, astonished, as Deming told them that if they adopted his teaching they would begin to capture world markets in a very few years. But Deming was no longer talking just statistics. Indeed, it is apparent from his 1950 diary that he often left the teaching of statistical quality control to assistants, while he taught the important concepts which he referred to as "the theory of a system, and co-operation". On one occasion he said, "The control chart is no substitute for the brain." Some might identify his teaching in Japan with what is today called TQM.

The Japanese realised that his ideas made a lot of sense and constituted a very different approach from that they were currently following. And, considering the situation, what did they have to lose? So they implemented Deming's concepts with an enthusiasm and dedication beyond Dr. Deming's expectation.

Deming's Message to the Japanese

Deming encouraged the Japanese to adopt a systematic approach to problem solving, which later became known as the Deming or Plan-Do-Check-Act **(PDCA) Cycle**. Deming, however, referred to it as the Shewhart Cycle, named after his teacher W. A. Shewhart [1931]. He subsequently replaced "Check" by "Study", as that word reflects the actual meaning more accurately. Therefore an alternative abbreviation for the Deming Cycle is **PDSA Cycle**.

Deming also pushed senior managers to become actively involved in their company's quality improvement programmes. His greatest contribution to the Japanese is the message regarding a typical business system (Figure 2.1). It explains that the consumers are the most important part of a production line. Meeting and exceeding the customers' requirements is the task that everyone within an organisation needs to accomplish. Furthermore, the management system has to enable everyone to be responsible for the quality of his output to his internal customers.

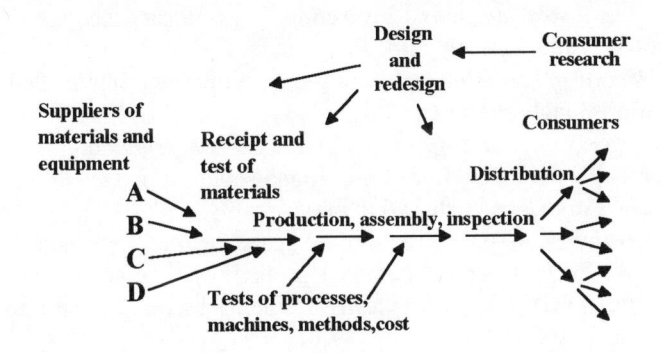

Figure 2.1 A typical business system [Deming, 1986]

Deming's Message to the West

"You do not have to do this. Survival is not compulsory."
                                            -- *W. Edwards Deming*

Deming's thinking in the late 1980's can best be expressed as Management by Positive Co-operation. He talks about the **New Climate** (organisational culture) which consists of three elements.
- *Joy in Work,*
- *Innovation, and*
- *Co-operation.*

He has referred to this New Climate as 'Win: Win', as opposed to the 'I Win: You Lose' attitude engendered by competition. In his seminars in America in the 80's, he spoke of the need for 'the total transformation of Western Style of Management'. *He produced his 14 Points for Management [Deming, 1989], in order to help people understand and implement the necessary transformation.* Deming said that adoption of and action on the 14 points are a signal that management intend to stay in business. They apply to both small and large organisations, and to service industries as well as to manufacturing.

Deming's 14 Points

1. *Create constancy of purpose for improvement of product and service.* Dr. Deming suggests a radical new definition of a company's role: A better way to make money is to stay in business and provide jobs through innovation, research, constant improvement and maintenance.
2. *Adopt the new philosophy.* For the new economic age, management need to take leadership for change into a 'learning organisation'. Furthermore, we need a new belief in which mistakes and negativism are unacceptable.
3. *Cease dependence on mass inspection.* Eliminate the need for mass inspection by building quality into the product.
4. *End awarding business on price.* Instead, aim at minimum total cost and move towards single suppliers.
5. *Improve constantly and forever the system of production and service.* Improvement is not a one-time effort. Management is obligated to continually look for ways to reduce waste and improve quality.
6. *Institute training.* Too often, workers have learned their job from other workers who have never been trained properly. They are forced to follow unintelligible instructions. They can't do their jobs well because no one tells them how to do so.
7. *Institute leadership.* The job of a supervisor is not to tell people what to do nor to punish them, but to lead. Leading consists of helping people to do a better job and to learn by objective methods.
8. *Drive out fear.* Many employees are afraid to ask questions or to take a position, even when they do not understand what their job is or what is right or wrong. They will continue to do things the wrong way, or not do them at all. The economic losses from fear are appalling. To assure better quality and productivity, it is necessary that people feel secure. "The only stupid question is the one that is not asked."
9. *Break down barriers between departments.* Often a company's departments or units are competing with each other or have goals that conflict. They do not work as a team, therefore they cannot solve or foresee problems. Even worse, one department's goal may cause trouble for another.
10. *Eliminate slogans, exhortations and numerical targets for the workforce.* These never help anybody do a good job. Let workers formulate their own slogans. Then they will be committed to the contents.
11. *Eliminate numerical quotas or work standards.* Quotas take into account only numbers, not quality or methods. They are usually a guarantee of inefficiency and high cost. A person, in order to hold a job, will try to meet a quota at any cost, including doing damage to his company.
12. *Remove barriers to taking pride in workmanship.* People are eager to do a good job and distressed when they cannot. Too often, misguided supervisors, faulty equipment and defective materials stand in the way of good performance. These barriers must be removed.

13. *Institute a vigorous programme of education.* Both management and the work force will have to be educated in the new knowledge and understanding, including teamwork and statistical techniques.
14. *Take action to accomplish the transformation.* It will require a special top management team with a plan of action to carry out the quality mission. Workers cannot do it on their own, nor can managers. A critical mass of people in the company must understand the 14 points.

Deming's System of Profound Knowledge

Before his death in December 1993, Deming summarised his 70 years' vision and experience and called it the *System of Profound Knowledge* [Deming, 1993]. Profound knowledge encompasses four interrelated dimensions [Neave, 1990]:

1. **Appreciation for a System**

This emphasises the need for managers to understand the relationships between functions and activities. Everyone should understand that the long term objective of an organisation is for everybody to win - employees, share holders, customers, suppliers, and the environment. Failure to accomplish the objective causes loss to everybody in the system.

However, there is no doubt that optimisation of a sub-system is easier to achieve than optimisation of the whole system. But it is also costly: it gives the impression of improvement yet, in reality, builds barriers which obstruct genuine progress. Optimisation of one part also harms other parts so that, overall, the change causes more harm than good.

Examples of sub-optimisation are: the destructive effects of grading in schools from the age of a toddler up to the secondary school and then all the way through to university; gold stars and prizes in school; the destructive effect of the so-called merit system; incentive pay; pay for performance; management by results; management by imposition of results; quotas.

Another form of sub-optimisation is caused by short-termism. Typical examples are:
- The monthly or half-yearly results. Make them look good. Defer till next month repairs, maintenance, orders for material. Aim for quick return on investment. High dividends now; never mind the future.
- Pension funds must be invested for high apparent return. This requirement leads to rapid movement of huge sums of money -- churning money out of one company into another.
- Ship everything on hand at the end of the month. Never mind its quality; mark it shipped, show it as accounts receivable.

Often a company's main aim is just to get a bigger slice of the pie than another company. The same holds with nations. If this is the sole or prime aim, the result is loss. The aim must be to bake a bigger pie, which brings more profit to both the company and its competitors. Everybody wins on that.

2. Knowledge of Statistical Theory

"We'll have to learn from the mistakes that others make. We can't live long enough to make them all ourselves."

-- *W. Edwards Deming*

This includes knowledge about variation, process capability, control charts, interactions and loss functions. All these need to be understood to accomplish effective leadership, teamwork, etc.

Firstly, there are two causes of variation, special and common. **Special causes** of variation in a product, process or service are those causes which prevent its performance from remaining constant. These special causes are often easily assigned: change of operator, shift, or procedure, for example. They can often be identified, and sometimes eliminated by local operators. On the other hand, **common causes** are those which emerge once the special causes have been eliminated. They are due to the design, or the operation of the process or system. Examples are poor design, inadequate equipment and procedure. They may be identified by the operators, but only management can eliminate common causes.

Deming believed that managers who lacked this understanding of variation, and confused the two types of variation could actually make matters worse. Furthermore he estimated that management was accountable for up to 98% of the potential improvement. Most losses are unknown, often unrecognised, not even suspected. There are two kinds of mistakes, both causing huge losses:

Mistake 1: *To react to any fault, complaint, mistake, breakdown, accident, shortage as if it comes from special causes when in fact it comes from the random variation due to **common causes**.* For example, a computer operator may try to adjust the set-up file of a computer without realising that the computer is limited by its hardware capacity.

Mistake 2: *To attribute to common causes any fault, complaint, mistake, breakdown, accident, shortage when actually it comes from **special causes**.* For example, a computer operator may think that a particular word processor is incapable of checking his spelling mistake without realising that there are better spell-check dictionaries for his job.

3. Theory of Knowledge

All plans require prediction based on past experience. An example of success cannot be successfully copied unless the theory is understood. For instance, there is no true value of any characteristic, state, or condition that is defined in terms of measurement or observation. There is a number that we get by carrying out a procedure. Change the procedure, and we are likely to get a different number.

By definition, there is no prediction in an unstable system; in a stable system, prediction is provided by control limits. In either case, there is considerable difference between the past and future when looking for what is best. Interpretation of data from a test or experiment is prediction. This prediction will depend on knowledge of the subject-matter, not just on statistical theory. Theory leads to prediction. Without prediction, experience and examples teach nothing.

Copying an example of success, unless understood with the aid of theory, may lead to disaster. Something that is excellent for somebody else, or in another place, may be destructive for us.

Operational objectives put communicable meanings into concepts. We need to know precisely what procedure to use in order to measure or judge something, and we need an unambiguous decision-rule to tell us how to act on the result obtained. Communication and negotiation (as between customer and supplier, between management and union, between countries) require optimisation of operational objectives.

There is no such thing as a fact based on an empirical observation. Any two people may have different ideas on what is important to know about any event, and hence what to record concerning anything which has happened. In this situation, there is no meaning for "Get the facts!"

4. Knowledge of Psychology

It is necessary to understand human interactions. Differences between people must be used for optimisation by leaders. People have intrinsic motivation to succeed in many areas. Extrinsic motivations in employment may smother intrinsic motivation. These include pay rises and performance grading. It has removed joy in work and in learning. We must give back to people intrinsic motivation for innovation, for improvement, for joy in work, for joy in learning.

We must also avoid over-justification. Over-justification comes from faulty systems of reward. Over-justification is resignation to outside forces. It could be monetary reward to somebody, or a prize for an act or achievement that he did for sheer pleasure and self-satisfaction. The result of reward under these conditions is to throttle repetition; the person will lose interest in such pursuits; he may never do it again. Monetary reward is often used as a way out for those managers who do not understand how to manage intrinsic motivation.

In the ITV-Central documentary in 1988 called *Doctor's Orders,* Dr Deming was asked why it was that the Japanese had paid attention to his teaching all those years ago while he continued to be ignored in America for a further 30 years. He answered: "No matter how old or how successful a Japanese manager is, he's always eager to learn. And he feels that he can put forth his best efforts because of that learning."

Deming would often emphasise the need to recreate the "yearning for learning" with which we are born. Children do not need to be paid, frightened, ranked and rated in order to learn their native language or to find out about the world around them. The founder of the British Deming Association, Dr. Henry

R. Neave [1995], pointed out in the First British Deming Memorial Lecture, 'The Man and his Message': "My final thought is this. If, in spite of his brilliance, his knowledge, his understanding, his experience, he was still so keen to *continue* learning in all humility from whatever the source, then what possible excuse there be for any of us who remain to ever *stop* learning?"

Just as Dr. Deming said in *Doctor's Orders:*

There is so much to learn: it's exciting, fun.'

■■

Mini-Case 2.1:

Zytec Implements Deming's 14 Points   [Ross, 1993]

Zytec designs and manufactures electronic power supplies and repairs power supplies and monitors. They were the 1991 Malcolm Baldrige National Quality Award winner for 1991. **Mini-Case 9.1** *has briefed how they managed to win the Award. This Mini-Case will focus deeper into their Leadership (Category 1), Strategic Quality Planning (Category 3), and Human Resource Utilisation (Category 4) to find out how they made use of Deming's 14 Points.*

MBNQA Category 1: Leadership

When Zytec started its quality improvement in January 1984, it chose Quality, Service, and Value as its focus. These developed into the company's Mission Statement as follows:

- *Zytec is a company that competes on value, it provides technical excellence in its products and believes in the importance of execution.*
- *We believe in a simple form and a lean staff, the importance of people as individuals, and the development of productive employees through training and capital investment.*
- *We focus on what we know best, thereby making a fair profit on current operations to meet our obligations and perpetuate our continued growth.*

To carry out this mission, Zytec's senior executives decided to embrace Dr. W. Edwards Deming's 14 Points for Management as the cornerstone of the company's quality improvement culture. They established the Deming Steering Committee to guide the Deming process and championed individual Deming Points, acted as advisors to the three Deming Implementation Teams, and developed the Zytec Total Quality Commitment statement.

Meetings were held with every Zytec employee to increase knowledge of Deming's Points, and many employees have attended half-day, two-day, and four-day Deming seminars. As a result, Deming's 14 Points for Management guide Zytec's actions, from Long Range Strategic Planning (LRSP) to employee

Chapter 2: The TQM Gurus' Ideas

empowerment to leadership. The effect of their compliance to the 14 Points is visible in their submission document for the MBNQA.

MBNQA Category 3: Strategic Quality Planning
Zytec's planning process takes the quality and service needs of customers and drives them through the organisation and to its suppliers. The process involves three steps:
- *Data is gathered.*
- *Goals are set by LRSP cross-functional teams.*
- *Detailed action plans to implement these goals are developed by department 'Management By Planning' teams.*

The Deming approach to setting objectives and developing plans for quality leadership requires that planning is based on data. Zytec collects this data by soliciting customer feedback, conducting market research, and benchmarking customers, suppliers, competitors, and industry leaders.

Because of the broad involvement and cross-functional development of the LRSP and management by planning, Zytec's direction for the future, both short- and long-range, has broad consensus and support. It exemplifies Dr. Deming's 14th Point: "Put everybody in the company to work to accomplish the transformation. The transformation is everybody's job."

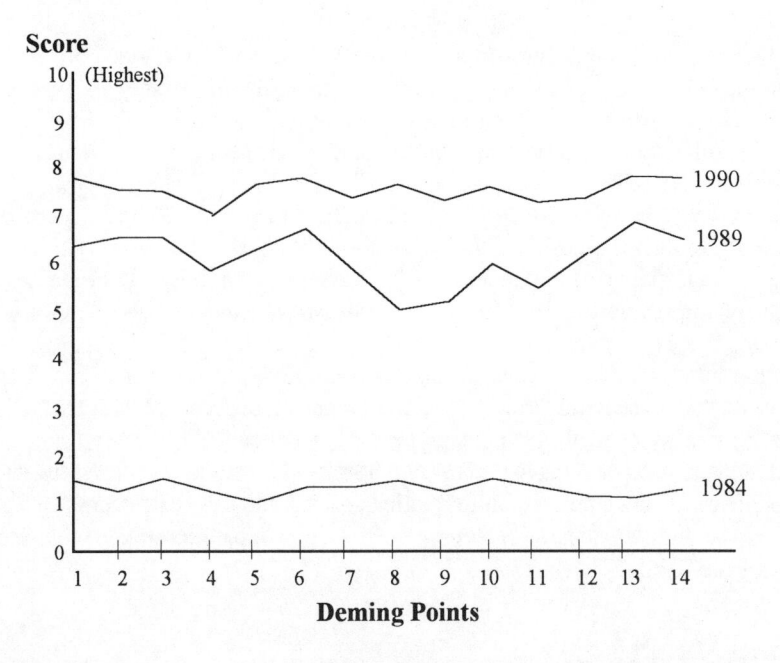

Figure 2.3 Effectiveness of Implementation of Deming's 14 Points by Zytec

MBNQA Category 4: Human Resource Utilisation
Human resource planning is guided by Dr. Deming's 14 Points and the LRSP. The results of a recent planning cycle provide an example of how these are translated into short- and long-term objectives:
- *Deming's Point 7: Institute leadership. The aim of leadership should be to help people and machines and gadgets do a better job. Leadership of management is in need of overhaul, as well as leadership of production workers.*
- *LRSP: Implement self-managed work groups in which employees make most day-to-day decisions while management focuses on coaching and process improvement.*

The results of training, involvement, and empowerment is that Zytec's employees believe Dr. Deming's 14 Points are more than vague guidelines. Since 1984, Zytec has surveyed employees seven times to gauge how effectively the company has implemented Deming's Points. As shown in Figure 2.3, Deming's Points provide an accurate description of the way Zytec conducts business.

■■

Joseph M. Juran

Like Deming, Dr. Joseph Juran is a charismatic figure of senior age, born in 1904. Juran started out as an engineer in 1924. In 1951 his first *Quality Control Handbook* was published and led him to international eminence. Chapter 1 of the book was titled *The Economics of Quality* and contained his now famous analogy to the costs of quality: 'there is gold in the mine'.

Juran was also invited to Japan by the JUSE, in the early 1950s. He arrived in 1954 and conducted seminars for top and middle-level executives. His lectures had a strong managerial flavour and focused on planning and organisational targets for improvement. He emphasised that *quality control should be conducted as an integral part of management control.*

Juran has had a varied career in management and his interest has gone beyond quality to the underlying principles common to all managerial activity. His 12 books on quality control and management have collectively been translated into 13 languages. He has received more than 30 medals, honorary fellowships, etc. in 12 countries for his contribution to quality. As was the case with Deming, among these is the highest decoration presented to a non-Japanese citizen, the Second Order of the Sacred Treasure.

Juran's Message

Juran developed the idea of **quality trilogy:** *quality planning, quality improvement and quality control.* These three aspects of company-wide strategic

quality planning are further broken down in Juran's 'Quality Planning Road Map', into following key elements:
Quality Planning
1. Identify who are the customers.
2. Determine the needs of those customers.
3. Translate those needs into our language.
4. Develop a product that can respond to those needs.
5. Optimise the product features so as to meet our needs and customer needs.

Quality Improvement
6. Develop a process which is able to produce the product.
7. Optimise the process.

Quality Control
8. Prove that the process can produce the product under operating conditions.
9. Transfer the process to Operations.

Juran concentrates not just on the end customer, but identifies other external and internal customers. His concept of quality is that everyone in the organisation must also consider the 'fitness for use' of the interim product at each stage of production/operation flow. He illustrates this idea about internal customers via the *Quality Spiral* [Juran, 1988]:

Customers ==> Product Development ==> Operations ==> Marketing

==> Customers ==> Further Product Development ==> etc.

Juran's work emphasises the need for *specialist knowledge and tools for successful conduct of the **quality function**.* He emphasises the need for continuous awareness of the customer in all functions. The mission of his recent teachings are to:
- Create an awareness of the quality crisis of the 1980s as highlighted by the NBC documentary "If Japan Can, Why Can't We?" (see Chapter 1).
- Establish a new approach to quality planning, and training.
- Assist companies to re-plan existing processes avoiding quality deficiencies
- Establish mastery within companies over the quality planning process thus avoiding the creation of new chronic problems.

Juran has found that the widespread move to raise quality awareness in the emerging *quality crisis* of the early 1980s is failing to change organisations' attitude towards quality. Whilst quality awareness was raised, it seldom resulted in 'doing it right first time' behaviour. Juran sees the failure as a result of the campaign's lack of planning and substance: "The recipe for action should consist of 90% substance and 10% exhortation, not the reverse." [Juran, 1988]. His formula for achieving success in a quality campaign is to:
1. Establish specific goals to be reached.
2. Establish plans for reaching the goals.

3. Assign clear responsibility for meeting the goals.
4. Base the rewards on results achieved.

Finally, in his "Last Word Conference" in 1994 [Abraham, 1995], Dr. Juran predicted for the future as:
- quality competition will intensify with multinationals and common markets
- there will be intense demands on suppliers
- ISO 9000 will sweep across the world
- Awards, e.g. Baldrige and European Quality Award, will supply intense stimulus and there will be a growth in awards world-wide

Philip Crosby

Philip Crosby, born in 1926, began his quality career as a reliability engineer. He later participated in the Martin missile experience that spawned the genesis of the zero defects movement, and then worked in quality management for ITT. He worked his way up within ITT and for fourteen years he was a Corporate Vice President and Director Quality of ITT, with world-wide responsibilities for quality.

In 1979 he published *Quality is Free,* which became a bestseller. In response to the interest shown in the book, he left ITT that year to set up Philip Crosby Associates Incorporated. At the Quality College, established in Florida, he started to teach organisations how to manage quality the way he advocated in his book.

Crosby published his second bestseller, *Quality Without Tears,* in 1974. He is also the author of *The Art of Getting Your own Sweet Way, More Things, The Eternally Successful Organisation and Leading, The Art of Becoming An Executive,* and many more.

Crosby's Message

Crosby's name is best known in relations to the concepts of **Do It Right First time** and **Zero Defects.** He considers traditional quality control, acceptable quality limits and waivers of sub-standard products to represent failure rather than assurance of success. Crosby therefore defines quality as conformance to the requirements which the company itself has established for its products based directly on its customers' needs. He believes that since most companies have organisations and systems that allow deviation from what is really required, manufacturing companies spend around 20% of their revenues doing things wrong and doing them over again. According to Crosby this can be 35% of operating expenses for service companies.

He does not believe that workers should take prime responsibility for poor quality; the reality, he says, is that you have to get management straight. In the Crosby scheme of things, management sets the tone on quality and workers follow their example; whilst employees are involved in operational difficulties and draw them to management's attention, the initiative comes from the top. Zero defects means that the company's objective is 'doing things right first time'. This will not prevent people from making mistakes, but will encourage everyone to improve continuously.

In the Crosby approach the Quality Improvement message is spread by creating a core of quality specialists within the company. There is strong emphasis on the top-down approach, since he believes that senior management is entirely responsible for quality.

The ultimate goal is to train all the staff and give them the tools for quality improvement, to apply the basic precept of Prevention Management in every area. This is aided by viewing all work as a process or series of actions conducted to produce a desired result. A process model can be used to ensure that clear requirements have been defined and understood by both the supplier and the customer. He also views quality improvement as an ongoing process since the work 'programme' implies a temporary situation. Crosby's Quality Improvement Process is based upon the

Four Absolutes of Quality Management

1. Quality is defined as conformance to requirements, not as 'goodness' or 'elegance'.
2. The system for causing quality is prevention, not appraisal.
3. The performance standard must be Zero Defects, not "that's close enough".
4. The measurement of quality is the Price of Nonconformance, not indices.

In *Quality is Free,* Crosby identifies additional quality-building tools, including the Quality Management Maturity Grid which enables a company to measure its present quality position. He also classifies **Cost of Quality** into three categories:

A. Prevention Costs -- The cost of all activities undertaken to prevent defects in design and development, purchasing, labour, and other aspects of beginning and creating a product or service. Also included are those preventive and measurement actions conducted during the business cycle. Specific items are:
- Design reviews
- Product qualification
- Drawing checking
- Engineering quality orientation
- Make Certain programme
- Supplier evaluations
- Supplier quality seminars
- Specification review
- Process capability studies
- Tool control
- Operation training
- Quality orientation
- Acceptance planning

- Zero Defects programme
- Quality Audits
- Preventive maintenance

B. Appraisal Costs -- These are costs incurred while conducting inspections, tests, and other planned evaluations used to determine whether produced hardware, software, or services conform to their requirements. Requirements include specifications from marketing and customers, as well as engineering documents and information pertaining to procedures and processes. All documents that describe the conformance of the product or service are relevant. Specific items are:
- Prototype inspection and test
- Production specification conformance analysis
- Supplier surveillance
- Receiving inspection and test
- Product acceptance
- Process control acceptance
- Packaging inspection
- Status measurement and reporting

C. Failure Costs -- Failure costs are associated with things that have been found not to conform or perform to the requirements, as well as the evaluation, disposition, and consumer-affairs aspects of such failures. Included are all materials and labour involved. Occasionally a figure must be included for lost customer credibility. Specific items are:
- Consumer affairs
- Redesign
- Engineering change order
- Purchasing change order
- Corrective action costs
- Rework
- Scrap
- Warranty
- Service after service
- Product liability

These specific items can be applied to both manufacturing and services organisations in general. To eliminate quality costs and to prove that quality is free, Crosby says that organisations must implement a quality management system to its fullest. Then it can turn what is sometimes considered a necessary evil into a profit centre.

The Japanese Quality Gurus' Ideas

Kaoru Ishikawa

Professor Ishikawa was born in 1915 and graduated in 1939 from the Engineering Department of Tokyo University. In 1947 he became an Assistant Professor at the University and was promoted to Professor in 1960. He was awarded the Deming Prize and the Industrial Standardisation Prize for his writings on Quality Control, and the Grant Award from the American Society for Quality Control in 1971 for his education programme on quality control. He died in April 1989.

Although the early origins of the famous Quality Circles can be traced back to the US in the 1950s, Professor Ishikawa is best known as a pioneer of the Quality Circle movement in Japan in the early 1960s (Chapter 6). This movement has now been re-exported to the West. In a speech to mark the 1000th quality circle convention in Japan in 1981, he described how his work took him in this direction: "I first considered how best to get grassroots workers to understand and practise quality control. The idea was to educate all people working at factories throughout the country but this was asking too much. Therefore I thought of educating factory foremen or on-the-spot leaders in the first place."

Ishikawa's Message

Ishikawa's biggest contribution is in simplifying statistical techniques for quality control in industry. At the simplest technical level, his work has emphasised good data collection and presentation, the use of Pareto Diagrams to prioritise quality improvements and Ishikawa Diagrams (see Chapter 6 for details).

Ishikawa sees the Cause-and-Effect Diagram (or Ishikawa Diagram), like other tools, as a device to assist groups or quality circles in quality improvement. As such, he emphasises open group communication as critical to the construction of the diagrams. Ishikawa diagrams are useful as systematic tools for finding, sorting out and documenting the causes of variation of quality in production and organising mutual relationships between them. Other techniques Ishikawa has emphasised include the seven Quality Control tools (see Chapter 6 for details).

Other than technical contributions to quality, Ishikawa is associated with the *Company-wide Quality Control (CWQC)* Movement that started in Japan during the period 1955--60 following the visits of Deming and Juran. Ishikawa sees the CWQC as implying that *quality does not only mean the quality of product, but also of after sales service, quality of management, the company itself and the human life.* The outcomes of such an approach are:

1. Product quality is improved and becomes uniform. Defects are reduced.
2. Reliability of goods is improved.
3. Cost is reduced.

4. Quantity of production is increased, and it becomes possible to make rational production schedules.
5. Wasteful work and rework are reduced.
6. Technique is established and improved.
7. Expenses for inspection and testing are reduced.
8. Contracts between vendor and vendee are rationalised.
9. The sales market is enlarged.
10. Better relationships are established between departments.
11. False data and reports are reduced.
12. Discussions are carried out more freely and democratically.
13. Meetings are operated more smoothly.
14. Repairs and installation of equipment and facilities are done more rationally.
15. Human relations are improved.

Shigeo Shingo

Shingo was born in Saga City, Japan in 1909. He graduated in Mechanical Engineering from Yamanashi Technical College in 1930. He was employed by the Taipei Railway Factory in Taiwan to improve their quality management. Subsequently he became a professional management consultant in 1945 with the Japan Management Association. It was in 1951, as Head of the Education Department, that he first heard of and applied statistical quality control. By 1954 he had investigated 300 companies. In 1955 he took charge of industrial engineering and factory improvement training at the Toyota Motor Company for both its employees and a hundred parts suppliers.

During the period 1956--58 at Mitsubishi Heavy Industries in Nagasaki, Shingo was responsible for reducing the time for the assembly of a 65,000 tons super-tanker from 4 months to 2 months. This established a new world record in shipbuilding, and the system spread to every shipyard in Japan. In 1959 he left the Japan Management Association and established the Institute of Management Improvement. In 1962 he started industrial engineering and plant-improvement training at Matsushita Electric Industrial Company. Training was introduced and conducted on a large scale, with some 7,000 workers involved.

It was in the period 1961--64 that Shingo extended the ideas of quality control to develop the Poka-Yoke, mistake-proofing or zero-defect concept. Subsequently the approach was successfully applied at various plants with over two years totally defect-free operation being established. In 1968 at Saga Ironworks he originated the Pre-Automation system which later spread throughout Japan. Also in that year he developed the Single Minute Exchange of Die (SMED) System at Toyota which is part of the Just-In-Time (JIT) System.

JIT is a production system whereby inputs are delivered to the production process just as they are needed. The benefit is that this eliminates costly inventory activities. More important is that whenever there is a quality problem, the whole production line is stopped until that problem is rectified. This enforces

Chapter 2: The TQM Gurus' Ideas

the whole plant to focus on solving the problem. Moreover, little re-work is required as the buffer stock in a JIT production system is minimal.

JIT requires tightly managed logistics to insure the timely arrival of inputs. Supplier relationships become a critical factor in the success of JIT production. Reliability and quality output are critical factors. JIT is part of the quality framework which includes rapid cycle-time to respond to customer needs. This requires a production design that allows flexibility. Recently, JIT finds applications in the services sector, apart from the manufacturing sector.

Shingo's first overseas study tour was in 1971. He visited Europe in 1973 at the invitation of Die-casting Associations in West Germany. He conducted practical training at Daimler Benz and Thurner. From 1975--79 he conducted training for the American company Federal Mogul on SMED and Non-stock Production. His first consultancy for an overseas firm was for Citroen in France in 1981. Other companies which he advised include many parts of Daihatsu, Yamaha, Mazda, Sharp, Fuji, Nippon, Hitachi, Sony and Olympus in Japan, and Peugeot in France. Shingo wrote more than 14 books. His book on the Toyota Production System and several others have been translated into English and other European languages.

Shingo's Message

In terms of quality, Shingo's paramount contribution was his development in the 1960s of Poka-Yoke and source inspection systems. These developed gradually as he realised that statistical quality control methods would not automatically reduce defects to zero.

The basic idea is to *stop the process whenever a defect occurs, define the cause and prevent the recurring source of the defect.* This is the principle of the JIT production system. No statistical sampling is therefore necessary. A key part of this procedure is that source inspection is employed as an active part of production to *identify errors before they become defects.* Error detection either stops production until the error is corrected, or it carries adjustment to prevent the error from becoming a defect. This occurs at every stage of the process by *monitoring potential error sources.* Thus defects are detected and corrected at source, rather than at a later stage.

Following a visit to Yamada Electric in 1961, Shingo started to introduce simple, mechanical devices into assembly operations, which prevented parts from being assembled incorrectly and immediately signalled when a worker had forgotten one of the parts. These mistake-proofing or 'Poka-Yoke' devices had the effect of reducing defects to zero.

In 1967 Shingo further refined his work by *introducing source inspections and improved Poka-Yoke systems which actually prevented the worker from making errors so that defects could not occur.* Associated advantages were that statistical sampling was no longer necessary, and that workers were more free to concentrate on more valuable activities such as identifying potential error sources.

Having learned about and made considerable use of statistical QC in his 40s, it was some 20 years later, in 1977, that Shingo observed that the Shizuoko plant of Matsushita's Washing Machine Division had succeeded continuously for one month with zero defects on a drain pipe assembly line with involvement of 23 workers. He realised that statistical QC is not needed for zero-defect operations. This was achieved principally through the installation of Poka-Yoke devices to correct defects and source inspection to prevent defects occurring. Together these techniques constitute *Zero Quality Control*, which, Shingo argues, can achieve what may have been impossible using statistical quality control methods.

Shingo advocated the practical application of zero defects by good engineering and process investigation, rather than slogans and exhortations that have been associated with the quality campaigns of many American and Western companies. Shingo, like Deming and Juran, argued that such American approaches of displaying defects statistics were misguiding and demoralising. Instead, the results of improvement should be announced and displayed.

Yoshio Kondo

Professor Kondo was born in 1924 and graduated in 1945 from the Kyoto University. He obtained his Doctorate in engineering and was promoted to Professor in 1961. He was awarded the Deming Prize in 1971 for his education programme on Quality Control. In 1987 he was conferred as an Emeritus Professor at the Kyoto University.

Kondo believes that quality is more compatible with human nature than cost and productivity because quality has effects on human life for longer period of time and is a common concern of both manufacturer and customers. To attain higher quality of products and service for customers, the motivation of employees is important.

Kondo's Message

Kondo emphasises the interrelationship between quality and people. He sees humanity as the essence of motivation. He endorses that human work should always include the following three components:
- Creativity -- the joy of thinking
- Physical activity -- the joy of working with sweat on the forehead
- Sociality -- the joy of sharing pleasure and pain with colleagues

He further points out that the elements of creativity and sociality are involved in company-wide quality control as well as physical activity, since the aim of CWQC is to ensure the superior quality of manufactured products and service through the stages of marketing, designing and manufacturing and, in so doing, to promote customer satisfaction. In other words, *there is no basic contradiction between CWQC activities and humanity.*

The major problems lie in the stages of designing the manufacturing process and evaluating the results of the work. When manufacturing is conducted only by standardising and simplifying the work and by separating planning from actual execution and when the results of the work are judged only in terms of money, how can we motivate workers by offering them meaningful jobs?

In his book *Human Motivation - A Key Factor for Management* published in 1989, Kondo advocates that making work more creative is important for motivation. He suggests four points of action in support of such a process:

1. **When giving work instruction, clarify the true aims of the work.**
   Instead of explaining clearly what the aim of a job is, people tend to concentrate on the methods and means to be used for achieving that aim. However, every job has an aim, and it goes without saying that achieving this aim is the most important thing. Aside from mandatory restrictions related to safety and quality assurance, information concerning means and methods should be given for reference only, and we should encourage people to devise their own best ways of achieving the objectives.

2. **See that people have a strong sense of responsibility towards their work.**
   This is related to the previous point. As we know well, human beings are often weak and irrational and tend to try to shift responsibility onto someone else when their work goes wrong, complaining or being evasive. It is, therefore, necessary to devise ways of nipping such excuses in the bud whenever they seem likely to appear. The 'mandatory objectives, optional means' approach described in Point 1 above serves this purpose, and techniques such as the stratification of data, the correction of data by mean value or by regression, and the application of the orthogonal principle in the design of experiments [Taguchi, 1986] are all effective devices for putting a stop to excuses.

3. **Give time for the creation of ideas.**
   Once people start feeling such a strong sense of responsibility, they will go back to the essence of the problem and think about it deeply. This will result in flashes of inspiration and the creation of new ideas. Excellent ideas are most easily generated during those times when we have pondered the problem deeply and have arrived at a detached, meditative state of mind. An ancient Chinese proverb tells us that this kind of time occurs when we are horseback riding, lying down and relaxing. The times at which ideas come most readily are different for every individual. The important thing is to give people the time to be creative.

4. **Nurture ideas and bring them to fruition.**
   New-born ideas created in this way are extremely fragile. If they are examined critically with the intention of picking them to pieces or squashing them down, it is very easy to obliterate them completely. However, to find out whether such ideas are really good or not, or to develop them in superior ways, they must be allowed to grow. There is no objection during this stage of growth to allowing an idea to change gradually from its original form into a better one. It is often said that the main enemies of new product development are found within the company itself. This means that people are more

concerned about going around stepping on new ideas than about encouraging their development. A new born idea is like a new-born baby, and raising it to maturity always requires someone to look after its interest and act as a loving parent. In most cases, those in positions of authority are the only ones who can play this role. In other words, managers should not go around throwing cold water on new ideas but should become their patrons and encourage their growth.

Kondo concludes that only by addressing all four points will it be possible for work to be reborn as a creative activity. If ideas are created and fostered, those concerned will come to feel a real sense of self-confidence. This is an extremely valuable experience from the standpoint of motivation.

Why are Quality Gurus' Ideas useful?

Quality Gurus' ideas are results of their life-time understanding and experience about quality. The ideas *have been tested by thousands of organisations world-wide before people recognised their originators as 'Gurus'*. The ideas are 'profound knowledge' that could be used by every individual and organisation, and their values are tremendous.

No quality improvement guru has all the answers. Each guru offers one unique idea to advance knowledge of solving quality-related problems. To solve real-life problems, one has to be able to understand, sequence, and synthesise the messages from different gurus to come to a useful conclusion.

Many companies may choose to follow the teachings of only one quality guru, assuming that listening to many might confuse them. Other companies may follow the teachings of all of the gurus and continuously synthesise ideas to make them applicable to their own operations.

Neave [1995] suggests that Deming's philosophy is a system -- thus everything in it is intimately related to much else. As he pointed out to the author in one communication: *"That is not to say I would expect any organisation to 'do' the whole of Deming! But there is need to understand the whole as a whole. The ice will be very thin otherwise. Once that understanding is there, the proliferation of ideas, techniques, strategies from other sources may be appraised with intelligence."*

In another instance, Bajaria [1995] suggests that *"We must learn to synthesise good ideas and make them integral to business practices"*. Thus, it is worthwhile to review the main ideas of the Quality Gurus.
1. W. Edwards Deming introduced concepts of variation to the Japanese and also a systematic approach to problem solving, which later became know as the Deming, PDCA or PDSA Cycle. Later in the West he concentrated on management issues and produced his famous 14 Points. He remained active till his death in 1993 and has summarised his 70 years experience in his System of Profound Knowledge.

2. Joseph M. Juran focused on Quality control as an integral part of management control in his lectures to the Japanese in the early 1950s. He believes that Quality does not happen by accident; it must be planned, and Quality Planning is part of the trilogy of planning, control and improvement. He warns that there are no shortcuts to quality.
3. Philip Crosby is best known in relation to the concepts of Do It Right First Time and Zero Defects. He based his quality improvement approach on Four Absolutes of Quality Management, the Cost of Quality and Quality Improvement Process.
4. Kaoru Ishikawa's three main contributions to quality were: 1) the simplification and spread of technical statistical tools (the 7 tools of Quality Control) as a unified system throughout all levels of Japanese companies, 2) his input to the company-wide Quality Movement, and 3) his input to the Quality Circle Movement.
5. Shigeo Shingo created the Poka-Yoke system to ensure zero-defects in production by preventive measures.
6. Yoshio Kondo identifies that quality is more compatible with human nature than cost and productivity. He developed a four point approach to motivation which makes it possible for work to be reborn as a creative activity.

How to Implement Some of the Quality Gurus' Ideas

In the process of introduction of TQM to an organisation it is advisable to take note of the unique messages of all the Quality Gurus. They have influenced most areas of TQM, as this chapter attempts to outline. The Bendell's seven point summary can be a guideline for getting the best out of the Quality Gurus [DTI, 1991].

1. Management commitment and employee awareness are essential from the early stages of TQM implementation. Deming's philosophy is possibly the most useful for encouraging these necessary attitudes.
2. The awareness should be backed up by facts and figures. Planning and data collection are important. Costs of Quality can be used to measure the progress of improvement. Juran has made the biggest impact in this area.
3. TQM programmes normally employ teamwork to facilitate improved communication and problem-solving. QCCs are particularly advocated by Ishikawa, and can be very successful if the rest of a TQM structure is in place.
4. Ishikawa advocated simple tools for problem-solving and improvement to be used by all employees.
5. There are also more technical tools to control industrial design and manufacturing. Shingo's work has been associated with successful Just-in-Time systems.
6. Management tools should be studied to achieve quality. These include the concepts of Company Wide Quality Control and Total Quality control associated with Ishikawa and Feigenbaum respectively.

7. In order to move from an inspection to a prevention culture, emphasis is placed on serving the internal customers and suppliers. This customer focus has been strongly stipulated by Juran's and Deming's recent teachings.

One priority that needs emphasising is that Deming's work is so challenging that it should deserve serious consideration. Finally, of vital importance is the need to develop a company-specific quality system. *It is likely that different companies will have different priorities and targets.* The Quality Gurus have an important contribution to make to TQM, but as has been pointed out by almost all Gurus, it can only be planned and driven by the senior management of the firm.

Chapter Summary:

- ▶ *The TQM Gurus are charismatic individuals whose concepts and approaches to quality within business, and life in general, have made major and lasting impact.*
- ▶ W. Edwards Deming (management philosophy) -- *Produces his 14 Points for Management in order to help people understand and implement the necessary transformation. Furthermore the Deming Cycle and the System of Profound Knowledge are important considerations in quality improvement.*
- ▶ Joseph M. Juran (quality trilogy) -- *Develops quality trilogy of quality planning, quality control and quality improvement.*
- ▶ Philip Crosby (zero defects and cost of quality) -- *His goal is to give all staff training and the tools of quality improvement, to apply the basic precept of Prevention Management in every area.*
- ▶ Kaoru Ishikawa (simple tools, QCC, company-wide quality) -- *Sees that quality does not only mean the quality of product, but also of after sales service, quality of management, the company itself and the human being.*
- ▶ Shigeo Shingo (Poka-Yoka/Fool-proofing) -- *Introduces source inspections and improved Poka-Yoke systems which actually prevented the worker from making errors so that defects could not occur.*
- ▶ Yoshio Kondo (four steps for making creative and quality work) -- *Sees that there is no basic contradiction between CWQC activities and humanity.*
- ▶ *TQM Gurus' ideas are useful because they have been tested by thousands of organisations world-wide before people recognised them as 'Gurus'.*
- ▶ *To implement TQM Gurus' ideas, it is likely that different companies will have different priorities and targets, since TQM is so fundamental and all-embracing. Deming's idea should deserve more in-depth considerations.*

Hands-on Exercise:

1. Consider the ideas of each of the six TQM Gurus, and highlight those which are applicable to your organisation.
2. Advise your Chief Executive on how you can manage the implementation of the TQM Gurus' ideas you identified above.

Chapter 3

The TQMEX Model

Topics:

- The need for a model in TQM
- What is TQMEX?
- Why do we need TQMEX?
- How to implement TQMEX
- The 4Cs of TQM
- The four pillars of quality
- What is Kaizen?

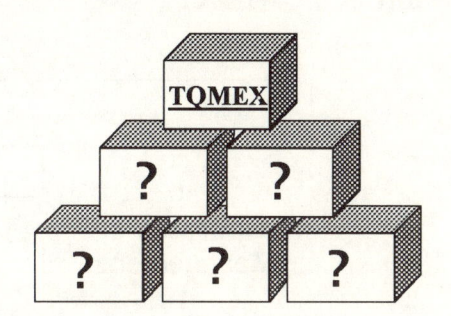

Introduction: The Need for a Model in TQM

At the century close, the creation of the global market, international orientation of management that sweeps national boundaries, introduction of new technologies, and shift towards customer focused strategies, make the competition stronger than ever. The criteria for success in this global, internationally oriented market have been changing rapidly. In order to expand business, enter new markets, and set realistic, competitive long-term objectives, *excellence* became an imperative. Management's effort has been directed towards discovering what makes a company excellent.

An approach, called *marketing concept* adopted by marketing-led organisations (for details see Chapter 5), is radical in its essence, because it emphasises the importance of relations with customers and role of quality in the search for excellence. To achieve excellence, companies must develop a corporate culture of treating people as their most important asset and provide a consistent level of high quality products and services in every market in which they operate.

Such an environment has supported the wide acceptance of Total Quality Management (TQM) which emerged recently as a new, challenging, marketable philosophy. It involves three spheres of changes in an organisation -- people, technology and structure. The role of top management in implementation of total quality is crucial and its input on people far-reaching. TQM, therefore, should be understood as management of the system through systems thinking, which means understanding all the elements in the company and putting them to work together towards the common goal. The TQMEX Model advocates an integrated approach in order to support the transition to systems management which is an ongoing process of continuous improvement that begins when the company commits itself to managing by quality. The Model illuminates the elements that form a base to the understanding of TQM philosophy and implementation of the process company-wide.

What is TQMEX?

```
                    ┌───────┐
                    │  5-S  │
         ↑          └───┬───┘
   Operations           ↓
   Management       ┌───────┐
                    │  BPR  │
         ↓          └───┬───┘
   ─────────────────────┼─────
         ↑              ↓
                    ┌───────┐
                    │ QCCs  │
                    └───┬───┘
                        ↓
                    ┌───────┐
    Quality         │  ISO  │
   Management       └───┬───┘
                        ↓
                    ┌───────┐
                    │  TPM  │
                    └───┬───┘
                        ↓
         ↓          ┌───────┐
                    │  TQM  │
                    └───────┘

5-S  = Seiri, Seiton, Seiso, Seiketsu, Shitsuke
BPR  = Business Process Re-engineering
QCCs = Quality Control Circles
ISO  = ISO 9001/2 Quality Management System
TPM  = Total Productive Maintenance
TQM  = Total Quality Management
```

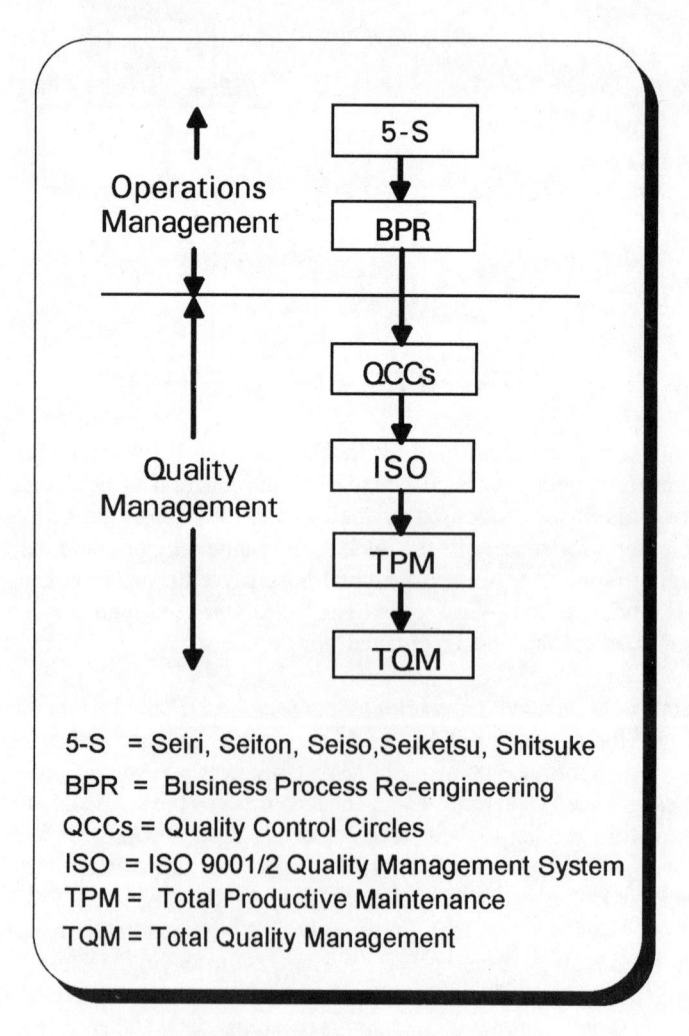

Figure 3.1 The TQMEX Model

In order to have a systematic approach to TQM, it is necessary to develop a conceptual model. Generally, a **model** is *a sequence of steps arranged logically to serve as a guideline for implementation of a process in order to achieve the*

Chapter 3: The TQMEX Model

ultimate goal. The model should be *simple, logical* and yet *comprehensive* enough for TQM implementation. It also has to sustain the changes in business environment of the new era. During my latest major overseas appointment as the first Quality Expert to the Malaysian Government in 1993, I was asked to develop a 5-year Quality Plan to contribute to the country's industrialisation programme. As a result of this assignment and my previous experience and research of some of the best TQM practices in Germany, Hong Kong, Japan, Korea, Malaysia, Taiwan, Singapore and UK, I proposed this TQM Model. The Model also reflects teachings of the contemporary quality gurus discussed in Chapter 2. The idea was to develop a universally applicable step-by-step guideline by including recognised practices in TQM (see Figure 3.1):

- Japanese 5-S Practice (5-S)
- Business Process Re-engineering (BPR)
- Quality Control Circles (QCCs)
- ISO 9001/2 Quality Management System (ISO)
- Total Productive Maintenance (TPM)

Each of these components will be introduced and defined in the following sections of this Chapter. They will be explained in greater depth in Chapters 4 - 9.

The Model's name is made up of **TQM** and **EX**cellence, indicating that *high standards and outstanding successes* are the goals. The ***TQMEX*** Model has been accepted by Dr. Tajuddin, the Director General of the Standards and Industrial Research Institute of Malaysia, as the 5-year Quality Plan and is being implemented in the 1995-1999 period. *He has also agreed to recruit another 20 consultants by the end of 1995 in order to reinforce the implementation team.*

TQMEX is a sequential model which is easy to remember and simple to implement. This is in line with the quality principle of Keep It Short and Simple **(KISS)**, although it is not simple to make a model simple (see Figure 3.2)!

Companies starting to implement TQM should follow TQMEX step-by-step. Companies which have already gone through some degree of improvement using some of the steps should review what have not been done and do it as their next step of improvement. In order to maximise your benefits from TQMEX, you have to start early too. The following are some of the challenges that everyone within each unit of a company should aim at using TQMEX:

1. To be profitable
2. To be 'recession proof'
3. To enjoy the results
4. To reinforce personal quality standards
5. To maintain customer confidence
6. To build customer loyalty
7. To improve customer satisfaction
8. To maintain corporate vitality
9. To use employees' creative energies
10. To develop a good reputation
11. To promote human dignity

12. To lower costs
13. To retain employees
14. To increase productivity
15. To contribute to society
16. To create a clear vision of the organisation
17. To improve technology
18. To solve problems effectively
19. To increase competitiveness
20. To develop internal co-operation

As a guideline, this model will facilitate the company's effort to achieve the standards required by TQM kitemarks world-wide. Chapter 9 is devoted to some of the most popular kitemarks today.

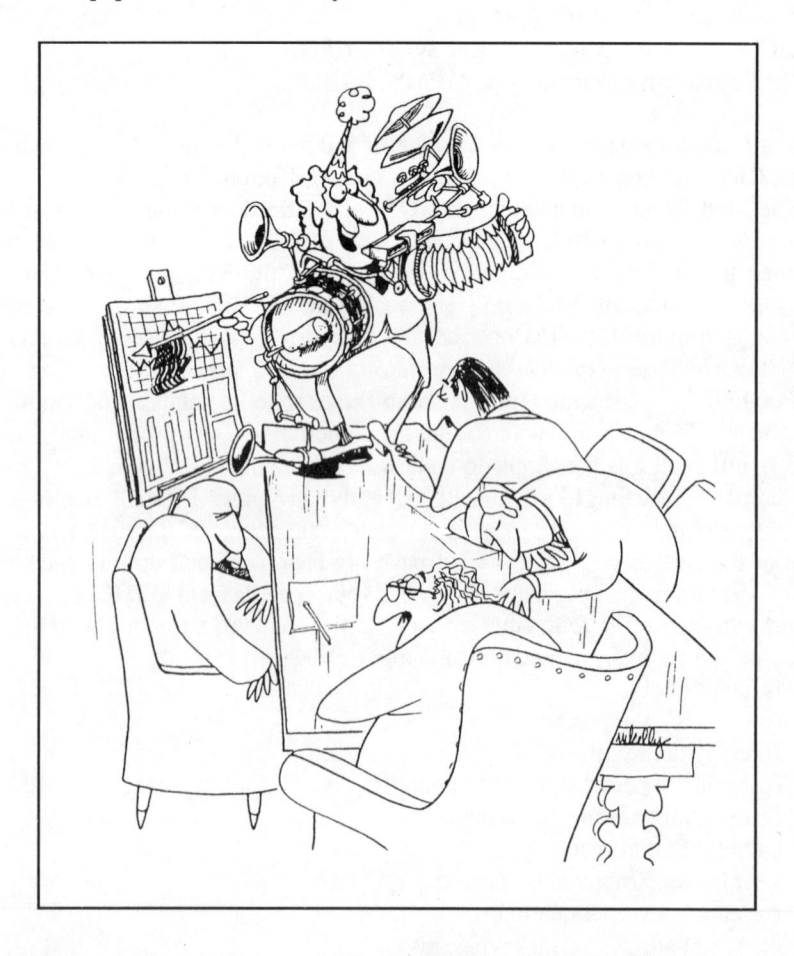

Figure 3.2 Keep It Short and Simple (KISS) is the fundamental of TQM [Migliore, 1992]

Japanese 5-S Practice (5-S)

The 5-S practice is a technique used to establish and maintain quality environment in an organisation. The name stands for five Japanese words: *Seiri, Seiton, Seiso, Seiketsu and Shitsuke* [Osada, 1991]. The English equivalent, their meanings and typical examples are shown in the following table:

JAPANESE	ENGLISH	MEANING	TYPICAL EXAMPLE
Seiri	Structurise	Organisation	Throw away rubbish
Seiton	Systemise	Neatness	30-second retrieval of a document
Seiso	Sanitise	Cleaning	Individual cleaning responsibility
Seiketsu	Standardise	Standardisation	Transparency of storage
Shitsuke	Self-discipline	Discipline	Do 5-S daily

The 5-S technique has been widely practised in Japan. Most Japanese 5-S practitioners consider 5-S useful not just for improving their physical environment, but also for improving their thinking processes too. Apparently the 5-S can help in all stratas of life. Many of the everyday problems could be solved through adoption of this practice. Unfortunately, unlike other quality tools and techniques, this basic but powerful technique for quality improvement has not been known to the western world. More detailed discussion of 5-S will be found in the next Chapter.

Business Process Re-engineering (BPR)

BPR is a management process used to *re-define the mission statement, analyse the critical success factors, re-design the organisational structure and re-engineer the critical processes in order to improve customer satisfaction.* BPR challenges managers to rethink their traditional methods of doing work and commit themselves to a customer-focused process. Many outstanding organisations have achieved and maintained their leadership through BPR [Oakland, 1995]. Companies using these techniques have reported significant bottom-line results, including better customer relations, reductions in cycle time to market, increased productivity, fewer defects/errors and increased profitability. BPR uses recognised techniques for improving business results and questions the effectiveness of the traditional organisational structure. Defining, measuring, analysing and re-engineering work processes to improve customer satisfaction pays off in many different ways.

Quality Control Circles (QCCs)

A QCC is *a small group of staff working together to contribute to the improvement of the enterprise, to respect humanity and to build a cheerful workgroup through the development of the staff's infinite potential.*

A quality control circle (QCC) *team of people usually coming from the same work area who voluntarily meet on a regular basis to identify, investigate, analyse and solve their work-related problems.*

It has been the Japanese experience that 95% of the problems in the workshop can be solved with simple quality control methods such as the 7 quality control tools [Ishikawa, 1986]. They are: Pareto diagrams, cause-and-effect diagrams, stratification, check sheets, histograms, scatter diagrams, and graphs & control charts. These tools will help QCCs to do brain-storming systematically and to analyse the problems critically. Then, through logical thinking and experience, most problems can be solved. QCCs and problem solving will be discussed in greater depth in Chapter 6.

The ISO 9001 Quality Management System (ISO)

The ISO 9000 series is a family of quality management and quality assurance standards developed by the International Organisation for Standardisation. It comprises of 17 different standards which are listed in **Annex 7.1** at the end of Chapter 7. Out of these 17 standards, only the ISO 9001, ISO 9002 and ISO 9003 are **quotable standards**, i.e., can be audited against. The others are **guidelines** only. ISO 9002 and ISO 9003 are sub-sets of ISO 9001. Most of the registered firms are registered under ISO 9001 or ISO 9002. Therefore, **ISO 9001:1994** will be used as the framework for quality management system in TQMEX (Chapter 7), together with ISO 9004-4:1993 as a guideline (Chapter 9).

ISO 9001 is the *Quality systems -- Model for quality assurance in design, development, production, installation and servicing.* It is the most comprehensive model of quality systems offered by ISO.

As quoted from the **Scope** of the ISO 9001:1994, this International Standard specifies quality system requirements for use where a supplier's capability to design and supply conforming product needs to be demonstrated. The requirements specified are aimed *primarily at achieving customer satisfaction by preventing nonconformity at all stages from design through to servicing.* This International Standard is applicable in situations when

a) design is required and the product requirements are stated principally in performance terms, or they need to be established; and
b) confidence in product conformance can be attained by adequate demonstration of a supplier's capabilities in design, development, production, installation and servicing.

ISO 9004-4:1993 is the Quality management and quality system elements -- Part 4: Guidelines for quality improvement. It gives suggestions for effective quality management, helps organisations in building their quality systems, so that they can develop quality improvement practices for TQM.

Total Productive Maintenance (TPM)

In 1971, the Japan Institute of Plant Maintenance (JIPM) defined TPM as *a system of maintenance covering the entire life of the equipment in every division including planning, manufacturing, and maintenance.* Because of its targeted achievement to increase productivity out of the equipment, the term TPM is sometimes known as Total Productivity Management (Chapter 8).

The JIPM runs the annual PM Excellence Award and they provide a checklist for companies applying for the award. There are 10 main items in the checklist:
- Policy and objectives of TPM
- Organisation and operation
- Small-group activities and autonomous maintenance
- Training
- Equipment maintenance
- Planning and management
- Equipment investment plans and maintenance prevention
- Production volumes, scheduling, quality, and cost
- Safety, sanitation, and environmental conservation
- Results and assessments.

Is TQMEX logical?

As Osada pointed out, 5-S is the key to total quality environment. Therefore, it should be the first step. BPR is concerned with re-defining and designing your business process in order to meet the needs of your customers effectively. It is more concerned with the business objectives and systems, and should follow as Step 2. QCCs are concerned with encouraging the employees to participate in continuous improvement and guide them through. They improve human resources capability to achieve the business objectives. Therefore, this should be Step 3. ISO 9000 is to develop a quality management system based on the good practices in the previous three steps. TPM is a result of applying 5-S to equipment based on a sound quality management system. In fact ISO 9001 requires procedures for process control and inspection and testing equipment which are part of TPM. Therefore TPM should be implemented in Step 5.

If the above five steps have been implemented successfully, the organisation is already very close towards achieving TQM. This is because by then the organisation will have had a good quality environment, well-defined business objectives and processes, a good quality culture, effective quality systems in place, and good equipment supports. It is a matter of choosing an appropriate TQM framework for further improvement (as will be discussed in Chapter 9). Finally, although the Quality Gurus (see Chapter 2) are not part of TQMEX, their ideas can be incorporated during TQMEX development, particularly in BPR, QCCs and ISO.

Pre-requisites for implementing TQMEX

In order to make a good base and proactive environment there are three pre-requisites that should be installed prior to TQMEX implementation. They are: the ten 'commandments', the four pillars, and the 4Cs of TQM.

The Ten TQM Commandments

In Chapter 1, it has been suggested that TQM should be integrated with corporate strategy. When formulating the corporate strategies, it is important to follow a set of guiding principles (commandments) to ensure that no important steps have been neglected in the process. TQM principles form the foundations for a successful implementation of TQM. They can be classified under ten major headings:

1. Approach -- Management Led
2. Method -- Prevention not Detection
3. Objective -- Total Customers Satisfaction
4. Measure -- The Costs of Quality
5. Standard -- Right First Time
6. Scope -- Ownership and Commitment
7. Theme -- Continuous Improvement
8. Ability -- Training and Education
9. Communication -- Co-operation and Teamwork
10. Reward -- Recognition and Pride

All these principles are equally important for the successful implementation of TQM and they should deserve the utmost attention by the management in the strategic planning and implementation processes.

The Four Pillars of TQM

The ten principles of TQM set the rules to be followed in acquiring TQM. On the other hand, there is a need for a systematic approach so that the ten principles can be bonded together smoothly. Oakland [1989] originated the idea of a 3-cornerstone model. The proposed 4-pillar model (Figure 3.3) brings the customer's requirement into the system. This makes the approach to TQM more complete. The additional pillar -- satisfying customers -- is vital because it explicitly addresses customers requirements. Without it TQM would have no objective. The interpretation of the four pillars (P1 to P4) are:

Chapter 3: The TQMEX Model 51

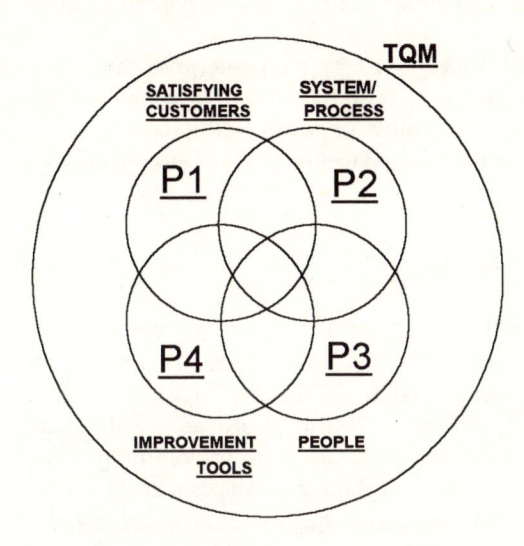

Figure 3.3 The Four Pillars of TQM

P1: Satisfying Customers -- The aim of TQM is not just to meet customer requirements (expressed needs). It is concerned about customer satisfaction (satisfying implied needs). Companies, such as Rover Cars, use the term "Extra-ordinary customer satisfaction" as their corporate mission. Customer requirements may include availability, delivery, reliability, maintainability and cost effectiveness, amongst many other features. The price of quality is therefore the continual examination of the requirements and your ability to meet them. Companies sometimes tend to neglect the needs of their employees. Only the smooth functioning of the internal customer-supplier chain will ensure that the quality is built into each stage of the business for the benefits of the external customers. The importance of satisfying the internal customers' needs can never be over-emphasised.

P2: System/Process -- For many firms, the first step in creating a total quality environment is likely to be the establishment of a quality management system such as ISO 9000 series , Ford Q-101, Rover RG-2000, etc. Establishing such a **system** is the initial building block. TQM relies on a effective quality management system which ensures that preventive measures are in place, and the culture of continuous improvement exists to enable the **processes** to deliver quality products and services. *Good processes will produce good products.*

P3: People -- It is vital for the management in a total quality organisation to capture the hearts and minds of everybody within the organisation, starting at the top and permeating via a chain of customer/supplier relationships throughout the whole organisation and beyond   (See Figure 1.5). Therefore management commitment, training, teamwork, leadership, motivation, empowerment, etc., would have a vital and complementary role to play in establishing a total quality environment.

P4: Improvement Tools -- There is no enterprise that cannot be improved. A vital part of TQM is to recognise the need for continuous improvement. The ISO 9004-4 Guideline for Quality Improvement should be a tangible help. The Guideline and a list of tools and techniques will be explained later in this Chapter.

The 4Cs of TQM

Implementation of TQM and improvement of a company's quality system heavily rely on people. Human dimension is inevitably incorporated in each of the stages of the TQMEX model, and is the driving force behind the philosophy of total quality in business as well as in life. It all starts with an individual's belief in actualisation through doing things with responsibility and pride, continuous learning, group work and contribution to common goals.

As much as any successful quality programme should be based on the Deming's cycle, the implementation of TQMEX also requires firm foundations in an organisation's human resources through a full application and integration of **4Cs of TQM: Commitment, Competence, Communication, and Continuous improvement.**

Commitment is the determined spirit of an Olympic swimmer who practises alone for hundreds and hundreds of pre-dawn hours. *Competence* is the inner knowing of a well-trained pilot who uses all available knowledge -- training, instruments, and intuition -- to make quick decisions. *Communication* is the critical personal contact and mutual agreement among managers and employees that makes work flow smoothly. *Continuous improvement (Kaizen* in Japanese*)* is the driving force towards excellence by each individual in an organisation. Therefore, in order to implement TQMEX, you need to prepare yourself first. The following four sections will explain each of the 4Cs respectively.

Commitment

Commitment to quality at work is defined as *a decisive personal or organisational choice to follow through on an agreed-upon plan of action.* Workers will be committed to quality to the extent that management is committed. For example, if the management is not committed to training the workers, they cannot produce the goods or services to the customers' requirement.

Everyone is committed to something to some extent. Our commitments vary according to their importance and our ability to meet them. Read this story:

Chapter 3: The TQMEX Model

> *A Cambridge University undergraduate who knew himself well, left this note on his door for his roommate: "Call me at seven. I absolutely have to get up at seven to study for my exam. Don't give up. Keep knocking until I answer." At the bottom he wrote: "Try again at nine."*

Do you think that the student is committed to his study? For a business to succeed in its commitment to quality, every employee must be committed to quality in every detail. This is necessary because everyone is adding value to the product or service he is responsible for. In order to satisfy the customers, every step of the process must be accomplished with commitment by the person performing it. An example of commitment to quality is given in the Mini-Case 3.1.

■■

Mini-Case 3.1:

On Commitment

Susan works as a keypunch operator typing coding sheets. She has been told, "Don't think, just type what you see." Susan knows she is typing obvious errors, such as 5 instead of a T. The programmer who receives the coding sheets corrects the errors and send them back to Susan for retyping, a time-consuming process.

One day Susan called the programmer and asked if he would stop by her desk to look at three obvious errors and correct them so she could type them correctly the first time, thus saving correction time later. While the programmer was still at his desk, Susan asked if he saw any other obvious errors that could be corrected now.

Does Susan have a commitment to quality ?   Yes ☐   No ☐
To what degree would you say she is committed? High ☐   Average ☐   Low ☐

■■

*After you have a good understanding of what Commitment means, it will be useful to do the **Self-Assessment Exercise 3.1** at the end of this Chapter for your own evaluation.*

Competence

Along with commitment, quality goals require actions and attitudes based on competence. Competence is based on knowledge. No matter what type of job you are doing, you must be competent to do it. You must possess certain specific measurable skills, sound education, good intuitive judgement, an ability to apply related knowledge to solve problems and a responsible attitude. Competence and quality work go hand in hand, because competent people make sure they successfully meet requirements.

Without competence, employees cannot build quality into a product or service. Successful TQMEX implementation raises morale and improve competence through education, teamwork, and incentive.

After you have a good understanding of what Competence means, it will be useful to do the Self-Assessment Exercise 3.2 at the end of this Chapter for your own evaluation.

Communication

The purpose of communication is to achieve mutual understanding. The definition of communication comes from the Latin word *communis* meaning, 'common'. Communication exist in most groups -- families, companies or among friends. Figure 3.4 illustrates how quality communication works. It shows the cycle of communication -- sender, message, receiver and feedback. Each column represents the important considerations for each of the four activities. Apart from their own responsibilities, the sender is responsible for the message and the receiver is responsible for the feedback.

Sender	Message	Receiver	Feedback
• Strategy • Direction • Goals	• Standards • Requirements • Expectations • Training	• Interpretation • Implementation • Participation	• Ideas • Suggestions • Problems • Challenges

Figure 3.4 Quality Communication Model

A common understanding and mutual agreement cannot take place with one-way messages. Too often we send a message and assume it is received and understood the way we intended. Here is an example:

An American tourist in a Madrid restaurant wanted to order steak and mushrooms. He spoke no Spanish; the waiter knew no English. The diner drew a picture of a mushroom and a cow. The waiter left the table and returned a few moments later with an umbrella and a ticket to the bullfights.

Chapter 3: The TQMEX Model

In the Quality Communication Model, both sides are responsible for success. For effective communication, the process is continuous until mutual understanding has been reached on the topic under discussion. However, in spite of good intentions, people often have problems getting the message across effectively. Between the *sender* and *receiver* something gets lost in the translation. The weak links in communication include an *unclear purpose, garbled messages, barriers* (such as a hidden agenda, cultural differences, jargon, etc.) and little or no *feedback*.

Suggestions for Communicating Quality

The message about quality needs to be communicated effectively by the management to everyone in the organisation. Here are some suggestions how this can be accomplished.
- Write policy statements
- Talk to customers
- Start a newsletter about quality
- Make a video about quality
- Create quality groups
- Management -- hold employee talk sessions
- Have suggestion boxes and forums
- Conduct status meetings
- Organise recognition parties for quality performance

An example of effective communication of quality in given in Mini-Case 3.2.

■■

Mini-Case 3.2:

Effective Communication of Quality   [Crosby, 1984]

A management team was concerned about the errors in computer programming. The team members felt that these put an unusually heavy burden on the mainframe computer, since it was being used primarily for troubleshooting. They estimated a cost of US$250,000 due to the problem of the software not being close enough to specification before troubleshooting began. Rather than just telling people about this, they borrowed ten brand-new Cadillacs and lined them all up in the front yard. Then they invited everybody out to "see what troubleshooting costs us." That made a big impression.

■■

*After you have a good understanding of what Communication means, it will be useful to do the **Self-Assessment Exercise 3.3** at the end of this Chapter. Check your answers with the ones suggested at the end of the book.*

Continuous Improvement (Kaizen)

"If you are content with the best you have done, you will never do the best you can do."

-- Anon.

Continuous improvement is a programme, a philosophy, and a strategy to improve the quality of goods and services of an organisation. Everything deteriorates with use. Maintenance programmes check the inevitable decline, but do not improve processes. Kaizen involves process changes. Renewal of a process can result in a major improvement in performance.

Team work (such as QCCs) and competence in problem solving are the foundation for *kaizen*. By adopting a structured methodology, the organisation ensures effective problem solving and decision making. By eliminating various root causes of problems in the process, variation decreases, thus increasing the quality of the output.

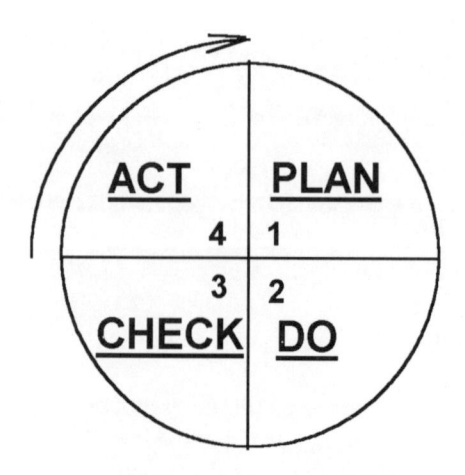

Figure 3.5 Deming's PDCA Cycle for Quality Improvement

At its elementary level, problem solving consists of the Deming's Plan-Do-Check-Act (PDCA) Cycle. This Cycle (Figure 3.5) becomes the basis for more sophisticated analysis. The sequence which drives the continuous improvement of processes is:

- Plan -- Data is collected, problem identified, and an improvement planned.
- Do -- A *pilot* project is done, or implemented.
- Check -- The results of the effort are observed and analysed against the plan.
- Act -- Then acted on, or deployed, throughout the organisation. At this stage the cycle starts again with planning an improvement.

In his later work, Deming replaced 'Check' with 'Study' because he wanted to emphasise the process of learning as more important than the limited action of checking -- inspection [Neave, 1990]. Consequently, the Deming Cycle became also known as the **PDSC Cycle.** Continuous improvement is an on-going and never-ending process of TQM. Therefore, a good understanding of this strategy and its practices are important for TQM. An example of kaizen is shown in Mini-Case 3.3.

■■■

Mini-Case 3.3:

Quality Improvement Storyboard at Komatsu Company [Vroman, 1994]

The storyboard is an elaborate problem-solving process. Japanese firms originated this comprehensive approach. Since then, many firms in Japan and America have initiated variations on this process. The intent is to insure that problems are fully investigated, solved, and implemented. One sure way of attaining a high level of complete problem solving is to have a work force trained and experienced in a comprehensive approach.

Kaizen results in constant diagnosis which, in turn, results in many problems to address. Establishing priorities, selecting candidates, and determining problem-solving approaches are decisions to be made. Counter-measures to problems are proposed that result in minimising sources of variation. When solutions are successful, permanent changes are proposed and implemented. The evaluation phase determines what problem area to take up next. The steps automatically support the firm's commitment to continuing improvement.

The Quality Improvement Storyboard originated with the Komatsu Company, a winner of the Deming Prize during the 1970s. Mr. Nogawa, president of Komatsu, named the process QI, for quality improvement. Florida Power & Light adapted the QI storyboard as the basis for their continuing improvement effort. They eventually become the first Deming Prize winner outside Japan.

■■■

Kaizen Strategy

There are always two external forces acting on the organisation: 1) the improvements made by the competition, and 2) the ever increasing pressure from customers and the market place. Company will fail if the people working in it have "Let's stop here", "Hold performance at this level", or "We have done enough" attitude (see Figure 3.6). Therefore it is important to build awareness of *kaizen* as a continuous process throughout the company. Management must incorporate this objective into their strategic planning and achieve it by a sound *kaizen strategy* which should be able to:

- Maintain and improve the working standard through small gradual improvements.
- Delineate responsibility for maintaining standards to the worker, with management's role being the improvement of standards.
- Produce a systems approach and problem solving tools that can be applied to realising this goal.
- Not a day should go by without some kind of improvement being made somewhere in the company.
- Use it as management tools within the TQM movement.

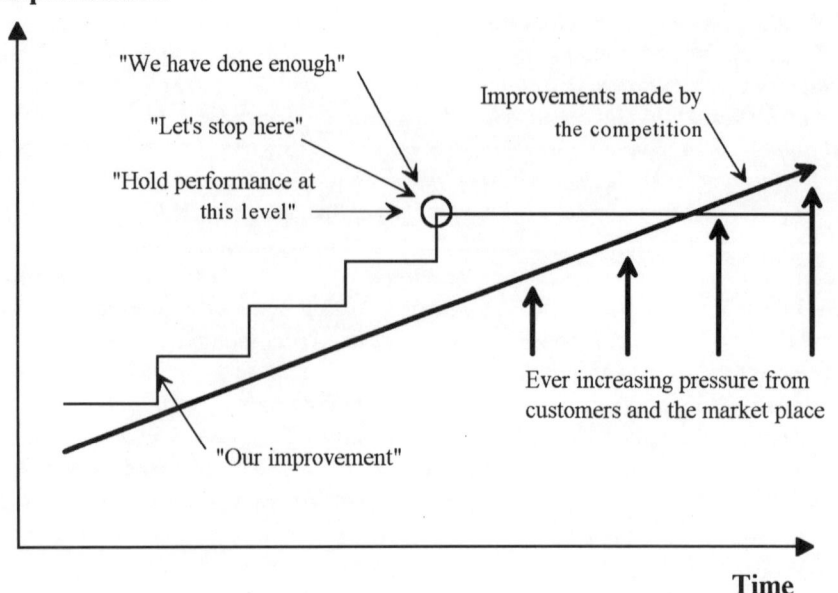

Figure 3.6 *Kaizen* should never stop

Chapter 3: The TQMEX Model

Kaizen Practices

Practising *Kaizen* requires the deployment of people, methods, tools and time. These elements fall into three categories: *kaizen* of management, *kaizen* of the group and *kaizen* of the individual. They are explained in details in Figure 3.7 below and should be followed closely by all parties concern.

	Kaizen of Management	*Kaizen* of the Group	*Kaizen* of the Individual
Tools	Seven statistical tools, Professional skills	Seven statistical tools, Common sense	Common sense, Seven statistical tools
Involvement	Managers and professionals	QCC members	Everybody
Target	Focus on systems and procedures	Within the same workplace	Within one's own work area
Duration	Until the completion of the *kaizen* project	Requires a few months to complete	Anytime
Achievements	As many as management chooses	A few per year	Many
Supporting system	Line and staff project team	Small-group activities, QCC, Suggestion system	Suggestion system
Implementation cost	Sometimes requires small investment	Mostly inexpensive	Inexpensive
Result	New system and facility improvement	Improved work procedure, Revision of standard	On-the-spot improvement
Booster	Improvement in managerial performance	Morale improvement, Participation, Learning experience	Morale improvement, *Kaizen* awareness, Self-development
Direction	Gradual and visible improvement, Marked upgrading of current status	Gradual and visible improvement	Gradual and visible improvement

Figure 3.7 *Kaizen* Practices

Chapter Summary:

▶ TQMEX is a systematic model encompassing the essential TQM methodologies to achieve excellence. It consists of the following elements:
  - *Japanese 5-S practice*
  - *Business process re-engineering*
  - *Quality control circles*
  - *ISO 9001/2*
  - *Total productive maintenance*
  - *Total quality management*

  The TQM EXcellence Model is based on sound TQM principles and logically presented.
▶ TQMEX is needed in order to satisfy customers consistently.
▶ The implementation of TQMEX is the subject of discussion in Chapter 10. However, in order to facilitate implementation, it is important to acquire by everyone in the organisation the 4Cs, i.e., Commitment, Competence, Communication, and Continuous Improvement (kaizen).
▶ The four pillars of TQM are: Satisfying customers, ISO 9000 quality management system, Improvement tools and people.
▶ Kaizen strategy is to maintain and improve the working standard through small gradual improvements. The practices include management-, group- and individual-oriented kaizen.

Self-Assessment Exercise 3.1: Rate your commitment

What is your level of commitment to quality in your organisation? Individuals make a difference, and many individuals working as a committed whole can revolutionise the quality and productivity of your organisation.

Listed below are personal and business situations that require a degree of commitment. Next to each situation put the appropriate number indicating which level of commitment you have to that item. Your answers are not right or wrong, they simply help you learn about yourself and where your commitment lies. Knowing yourself is the first step in making changes.

1. Unwavering
2. Diligent
3. Casual
4. When I am in the mood

____ Learning
____ Helping others
____ Customers
____ Personal quality standards

Chapter 3: The TQMEX Model 61

____ Enjoying life
____ Family
____ Organisational goals
____ Doing my best

Self-Assessment Exercise 3.2: Prioritise your Competence

Listed below are areas of personal and professional competence. Prioritise the list giving '1' to the highest priority requiring improvement, '2' the next priority, and so on. Start improving them with the help of your upper management, family, or friends if necessary.

TASK	PRIORITY
Job skills	_____
Education	_____
Job experience	_____
Communication skill	_____
Problem solving	_____
Decision making	_____
Hobbies	_____
Organisation	_____
Neatness	_____
Cleanliness	_____
Standardise work methods	_____
Self-discipline	_____
Others:	_____
	_____

Self-Assessment Exercise 3.3: Rate Your Communication

Quality is largely determined by how effectively you can communicate in your personal and work life. You can communicate more effectively when you understand the expectations in communication and set useable standards for yourself.

Answer the questions below to clarify your communication expectations and to set your communication standards.

1. When you are the sender of information, what do you need from the receiver in order to know you were understood accurately?
   _____
   _____

2. What can you do to get what you need?
   _____
   _____

3. How can you meet other peoples' expectations in a communication situation?
   _____
   _____

4. What is a reasonable standard (or goal) to get for increasing the quality and effectiveness of your present communication?
   _____
   _____

5. How will you measure your standard?
   _____
   _____

6. What are some methods of improving communication about the quality programme in your organisation?
   _____
   _____

Hands-on Exercise:

1. Consider TQMEX. Is there other TQM methodology which you think should be included?

2. How does the TQMEX fit into the strategy and operations of your company?

3. How would you prepare yourself for TQMEX implementation?

Chapter 4

Japanese 5-S Practice

Topics:

- What is the 5-S?
- The 5-S in details
- The 5-S in the office
- Why is the 5-S useful?
- How to implement the 5-S

Introduction: What is the 5-S?

Japanese factories are well-known for their cleanliness and orderliness. This results from their ability to instil a sense of responsibility and discipline into their workers, particularly at the plant level. The logic behind the 5-S practices (see Chapter 3 for definition) is that organisation, neatness, cleanliness, standardisation and discipline at the workplace are basic requirements for producing high quality products and services, with little or no waste, and with high productivity.

Surprisingly, this powerful quality tool has been a secret to the West. The western world has just recently recognised the significance of the 5-S practice although there are indications that some companies have included some aspects of the 5-S in their routines without being aware of its existence as a formalised technique. There are many examples of successful implementation of some principles of the 5-S, especially in the service sector organisations, such as fast-food restaurants, supermarkets, hotels, libraries, and leisure centres.

The difference between the Japanese and western approach lies mostly in the degree of employee involvement. By formalising the technique, the Japanese established the framework which enabled them to successfully convey the message across the organisation, achieve total participation and systematically implement the practice. The 5-S has become the way of doing business, not only to impress the customers but to establish effective quality processes as prerequisites for good products and services. Mini-Case 4.2 illustrates how a daily 5-minute 5-S drill can make a difference to the work environment.

The 5-S in Details

The following sections will explain each of the constituents of the 5-S practice in appropriate depth to enable practitioners to get the maximum benefit from its implementation, yet not making it too complicated to understand. While contemplating each of the 5-S aspects due reference should be made to the **5-S Audit Worksheet** in **Annex 4.1** at the end of this Chapter.

Whilst the 5-S Audit Worksheet intends to cover most of the potential areas for improvement, it is not limited to the points in the table only. Depending on the need of your organisation, there may be more important points which you should add to the table at your own discretion. This principle applies to all of the five categories of the 5-S. Some photos of good and bad examples of the 5-S practices are also included as additional guidelines for the 5-S audit.

What is Organisation (Seiri)?

Organisation is about *separating the things which are necessary for the job from those that are not and keeping the number of the necessary ones as low as possible and at a convenient location.*

In those times when the land was poor and supply of goods was naturally restricted, people would hold on to the least little thing because it seemed such a waste, almost a sin to throw anything away. Yet today, when there is an abundance of goods, services, and information, sorting through these things has almost become the art or rare skill. By looking at information alone, there is a whole new career field called *information management* that does nothing but sort through information and organise it. It is important to save things, but it is just as important to throw things out (see Mini-Case 4.1). And most important of all is knowing what to discard, what to save, and how to save things so that they can be accessed later.

■■

Mini-Case 4.1:

My Experience in Throwing Things Away

My working day at the University starts with taking the mail out of my 'pigeon hole'. I glance through it in the lift lobby and inside the lift on the way to my

Chapter 4: Japanese 5-S Practice

office. As I am walking, I separate the things I want to keep from the things I do not want to keep. And then the things that failed to pass this initial screening get dumped in the wastebasket. Doing this, I often find that by the time I am at my desk I have only 20% of the original lot left, half of which I have to take action about. One annoying experience I had was that once I received a nicely printed and collated set of 'Invigilators Timetable' of about 50 pages on coloured paper. I had to spend 10 minutes to go through it and to find out that my name only appeared in one row. So, besides throwing the set straight into the bin, I made a phone call to the Academic Registry advising them of their 'waste'.

■■

Stratification Management

The art of organisation is in stratification management. It involves deciding how important something is and then reducing the non-essential inventory. At the same time stratification management ensures that the essential things are close at hand for maximum efficiency. Thus the key to good stratification management is the ability to make these decisions about usage frequency and to ensure that things are in their proper places. It is just as important to have the things you do not need far from hand as it is to have the things you do need close at hand. It is just as important to be able to throw out a broken or defective part as it is to be able to fix it. A summary of organising things is shown in Figure 4.1.

Usage	Degree of Need (Frequency of Use)	Storage Method (Stratification)
Low	• Things you have not used in the past year • Things you have only used once in the last 6-12 months	• Throw them out • Store at a distance
Average	• Things you have only used once in the last 2-6 months, Things used more than once a month	• Store in central place in the workplace

High	• Things used once a week,
	   Things used every day,
	   Things used hourly
		• Store near the work site or carry by the person

Figure 4.1 The way to organise things according to usage

Differentiation between Need and Want

Once stratification and classification are done, you are in a position to decide what you want to do with things that you do not use more than once a year. Save them or throw them away? If you decide to save them, how much of them do you need to save? It is safe to assume you need less of something the less frequently you use it. And whenever you do this kind of major housecleaning you will find loads of junk you do not need. This is a never-ending process not only because things tend to collect, but also because it is very difficult to distinguish between what you do and what you do not need. Many people do confuse between **need** and **want**. Most of us have a colleague who has collected 60 trade magazines on his shelf over the last five years and claims that he *wants* to keep all of them for reference. If you ask him "Which of the 60 magazines do you *need*?", he would probably scratch his head and say "I do not know what you are talking about." However, if you formulate your question the following way: "Which of these 60 magazines have you not been touching over the last year?", you should not be surprised to find that the answer is 50.

Most people tend to err in the beginning, on the conservative side of saving things 'just in case'. But it is crucial that management make a decision. Is it needed? If not, get rid of it. If yes, how much of it is needed? Get rid of the rest. If something is borrowed, return it to its owner.

One-is-Best
Apart from throwing away rubbish, other aspects of organisation is shown in the 5-S Audit Worksheet (1.1 to 1.10). It is worthwhile to emphasise the importance of a principle of organisation called 'one-is-best'. Examples of application include: one set of tools/stationery, one page form/memo, one day processing (Figure 4.2), one stop service for customer and one location file (including local area network server for file sharing). In particular for 'one day processing', there is an ancient Chinese saying *"Let today's work belong to today"*. There is a lot of virtue in this saying and it requires a combined effort of organisation and self-discipline.

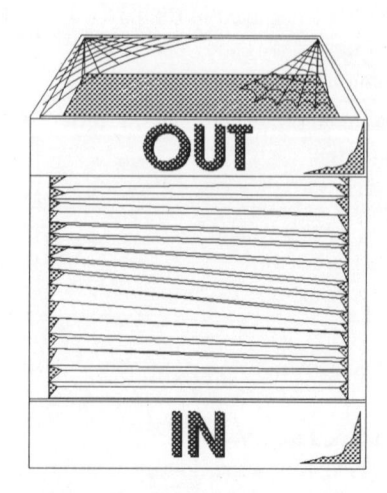

Figure 4.2 A Bad Example of One-is-Best --
Not complying with "One-day Processing"

Chapter 4: Japanese 5-S Practice 67

Photo 4.1 Good Example of Organisation -- Storage of parts and documents (1.5)

Photo 4.2 Bad Example of Organisation -- Throw away rubbish (1.1)

Two typical examples of good and bad organisations are shown in Photo 4.1 and Photo 4.2 respectively. The bracketed number refers to the number in the 5-S Audit Worksheet at the end of this Chapter.

What is Neatness (Seiton)?

Neatness is a study of efficiency. It is a question of how quickly you can get the things you need and how quickly you can put them away. Just making an arbitrary decision on where things go is not going to make you any faster. Instead, you have to analyse why getting things out and putting them away takes so long. You have to study this for both the people using the things frequently and those who seldom use them. You have to devise a system that everyone can understand. There are four steps in achieving neatness:

Step 1: Analyse the status quo
Start by analysing how people get things out and put them away, and why it takes so long. This is especially important in workplaces where a lot of different tools and materials are used, because time spent getting things out and putting them away is time lost. For example, if a person gets something out or puts something away 200 times a day and each time takes 30 seconds, you are talking about 100 minutes a day. If the average time could be reduced to 10 seconds, more than an hour could be saved.

If you are dealing with small-lot production and quick turnaround times, every second counts. A minute spent getting something out and putting it away could be fatal. Typical problems in retrieving things are:
1. Not knowing what things are called
2. Not sure where things are kept
3. Storage site far away
4. Storage sites scattered all around
5. Repeated trips
6. Hard to find because many things are there
7. Not labelled
8. Not there, but not clear whether it is finished or somebody is using it
9. Unclear if spare parts exist (no ledger and nowhere to ask)
10. One brought was defective
11. Hard to get out
12. Too big to carry
13. Need to set or assemble
14. Too heavy to carry
15. No gangway to transport

Step 2: Decide where things belong
The second step is to decide where things belong. Of course, there need to be criteria for deciding this, because the absence of criteria and any pattern will

Chapter 4: Japanese 5-S Practice

make it impossible for people to remember where things are supposed to be and will mean that it takes that much longer for them to put things back or to get them out. Yet there are many possibilities, and selecting the one that is best for you will take some study.

It often happens that an object can have two names: its real name and what everybody calls it. In such cases, make a decision which one you are going to use and stick to it. It only confuses people to have two names for the same object. On the other hand, during stock-take, you may find out that there are many things that do not have a name. There may be times when two different things have the same name, even when there are minor differences between them. You should rectify all these problems as soon as possible.

Step 3: Decide how things should be put away

The third step is to decide how things should be put away. This is critical to functional storage. For example, files and tools should be put away so that they are easy to find and easy to access. Storage has to be done with retrieval in mind.

Having a name for everything is not yet sufficient. Things must have a location, just like everybody would have a home. It is in fact quite amazing to send a letter through the world-wide mailing system. As long as you put on the right address, it will go to its destination, anywhere in the world. The principle is very simple -- there is a name on your letter and then there is a name on the location which matches with that on the letter. Therefore in doing your 'neatness', it is imperative that every object should have both a name and a location.

In assigning storage space, designate not only the location, but even the shelf. Decide where things should be, and make sure that they are at their home. This is crucial. When the storage location is on the tool and the tool's name is on the storage location, you know you are doing it right. The following procedures should be adhered to:

1. Everything should have a name
2. A place for everything and everything in its place
    -- No more homeless items
    -- Even if someone is just using something temporarily, it should be clear where it is
3. Quick identification
    -- Arrows and lamps
    -- Frequent-use items to be retrieved easily
4. Safe storage
    -- Heavy things on the bottom
    -- Heavy things on dollies
    -- Benches and ladders
5. Height considerations
    -- Knee to shoulder height most convenient

Step 4: Obey the put-away rules
The last step is to obey the rules. This means always putting things back where they belong. It sounds simple, and it is as if you would be doing it. It is just doing it that is difficult. Whether or not this is done will determine whether or not organisation and neatness succeeds. At the same time, inventory management is important to see that you do not run out of parts or products. In order to achieve this, the rules are:
1. Out of stock
    -- Decide on minimum stock level
    -- Indicate that more are on order
2. Somebody is using it
    -- Have an indication of who is using it and when they will return it
3. Lost
    -- Decide how many there should be
    -- Draw a shadow outline indicating clearly what is missing

Neatness for Notices and Signs

Since notices, posters and signs are common means of communication and they also project an image of how neat a company is, they should deserve special attention. The following guiding principles apply:
1. Do not just put them up anywhere. Designate special places for them and stick to the designations.
2. Every noticeboard should have a clear label. Large noticeboards should be divided into zones, with clear label indicating such headings as 'Urgent', 'General', 'External', 'Staff Communication', etc.
3. Be sure to indicate how long they are going to be up. Nothing should go up without an indication of when it comes down. This is necessary because not only are out-of-date posters waste of space, but they are up so long that nobody pays any attention to them anymore -- people get into the habit of not paying attention to any notices, even new ones.
4. Tape should be peeled off in such a way that the wall painting is not damaged afterwards.
5. Notices, posters and signs should be aligned neatly along the top.
6. Poster paper should be at least A3 to A2 sizes with decent size lettering.
7. Posters that are going to be hanging down from the ceiling should be carefully measured first to make sure they are not going to be in the way.
8. Posters should be put up firmly so that they do not blow around or fall down in the breeze when windows are open or when people walk by.
9. The messages may be hand-written, but they should be neat and legible. It is a good idea to use computer printed fonts.
10. Stands and signs are often a distraction. Heights should be carefully determined and placement should be carefully considered so that people can see what the signs are all about.

Chapter 4: Japanese 5-S Practice 71

Photo 4.3 Good Example of Neatness -- 30-second retrieval of files (2.2)

Photo 4.4 Bad Example of Neatness -- Everything should have a 'home' (2.1)

What is Cleaning (Seiso)?

'Everyone is a Janitor' -- Cleaning should be done by everyone in the organisation, from the managing director to the cleaner. This is why in Japan, they do not need street cleaners in residential areas. Every family is responsible for cleaning the pavement in front of their houses. Therefore, what they need are rubbish collectors. The Japanese believe that while they are doing cleaning, they are cleaning their minds, too. If you have done your annual cleaning at home before the New Year, you would probably have this feeling of freshness.

There are even companies that have taken steps towards putting **little gardens** in their workplace rest-areas as hygiene has ramifications well beyond the factories and offices to the surrounding environment. The mottoes for cleaning are:
- I will not get things dirty.
- I will not spill.
- I will not scatter things around.
- I will clean things right away.
- I will rewrite things that have got erased.
- I will tape up things that have come down.

An orderly progression of cleaning in the factory environment by piece of equipment and by location will often identify causes of various problems in the production process, such as:
- Dirty air-conditioning filters lead to defects in printing.
- Filings in the conveyance chutes lead to scratching.
- Scraps in the die leading to faulty pressings.
- Things fall off the equipment and get into the products.
- Things get dented or bent in conveyance.
- Filings and other particles contaminate the resin.
- Dirty coolant leads to clogging.
- Dust and other substances ruin the painting process.
- Bad connections are made because the electrical contacts are dirty.
- Fires are caused because garbage short-circuited the electrical equipment.
- Computer always plays up because dirt is accumulated inside

In an office or a factory, you might start by graphing out the individual areas of responsibility. In doing this, it is important that all assignments be absolutely clear and that there is no undefined, unallocated, or grey areas. Unless each and every person takes these admonitions to heart and accepts personal responsibility, you are not going to get anywhere.

Chapter 4: Japanese 5-S Practice 73

Photo 4.5 Good Example of Cleaning -- Sparkling clean campaign (3.3)

Photo 4.6 Bad Example of Cleaning -- Individual cleaning responsibility (3.1)

What is Standardisation (Seiketsu)?

Standardisation means continually and repeatedly maintaining your organisation, neatness and cleaning. As such, it embraces both personal cleanliness and the cleanliness of the environment. The emphasis here is on *visual management* and 5-S standardisation. Innovation and total visual management are used to attain and maintain standardised conditions so that you can always act quickly.

Visual Management
Visual management has recently come into the limelight as an effective means of continuous improvement. It has been used for production, quality, safety, and customer services. Colour management has also come in for considerable attention lately. This has been used not only for colour-coding, but also to create a more pleasant work environment. There are more and more workers opting for white and other light-coloured clothes. Because these clothes show the dirt quickly, they provide a good indicator of how clean the workplace is. They highlight the need for cleaning.

One effective method of visual management is to put up appropriate labels. Examples are:
- Lubricating oil label -- Indicate the type, grade, colour, and where it is for.
- Annual inspection label -- Should be attached to all equipment.
- Temperature label -- Indicate abnormality or overheat.
- Responsibility label -- Show who is responsible for what.
- Identification label -- Tell people what things are.
- Safety label -- Remind people of special safety considerations.
- Zone label on meters -- Normal zone and danger zone should be differentiated by different colours.
- OK mark -- After things have been inspected, an OK mark will tell others that the part is acceptable.
- Position mark -- Put little position marks for where things go. Place footprints where people should stand. Place marks on the floor to indicate danger zones. Place lines to indicate where things are supposed to stop. Put up lots of visual clues so that everybody will be able to see what is happening and to anticipate what will happen next.

From the overall human perspective, hygiene has ramifications well beyond the factory to the surrounding environment. The 5-S is applied to oil mist, filings, noise, thinner oils, toxic substances, and other sources of pollution found in the workplace. There are even companies that have put little gardens in their workplace rest-areas, calling them the *Factory Gardens,* an example of which is the Oriental Motors in Japan.

Transparency
Another important consideration for standardisation is 'transparency'. In most factories and offices, tools and files are put in lockers, on closed shelves, and

under covers to be off sight. Just like sweeping things under the carpet this is known as "out of sight, out of mind" practice. Those closed spaces are often among the most disorderly places, because they are not a constant eyesore. So it will be a good idea to take the wraps off these messes. Make the covers transparent. If you must have metal panels, put inspection windows in them. Make it so that everybody can see what is stored and how good (or bad) things look.

Visualising Conditions
Many places have little ribbons on the fans so you can see the breeze. Sometimes, this method is called 'visualising conditions'. As a variation of this, some plant maintenance people put windows and plastic strips in some of the drain pipes so that other people can see the effluent flowing. There are many other things you can do to help people visualise a process.

Trouble Maps
When there are problems, you can show them on a map of the workplace. Just as many sales departments have pins in maps to show where their people are, you can also have pins to show problems, emergency exits, fire-fighting equipment, and other important locations. Put the maps where they are visible to everyone. A trouble map can also be used to indicate those workplaces and processes that are trouble-free.

Quantification
By constantly measuring things, quantifying the results, and analysing the data statistically, you can quickly identify the limits to management and spot deviations before they become major headaches.

What is Discipline (Shitsuke)?

Discipline means instilling the ability of doing things the way they are supposed to be done. The emphasis here is on creating a workplace with good habits. By teaching everyone what needs to be done and having everyone practising it bad habits are broken and good ones are formed. This process helps people form habits of making and following the rules.

The word *shitsuke* originally comes from the guiding stitches that are done before a garment is properly sewn. If accepted that way, discipline is an underlying tool in making life go smoother. It is recognised by the Japanese as the minimum the society needs in order to function properly.

Self-discipline is important because it reaches beyond discipline. If a person is 'disciplined' to do something at one time there is a chance that he may not be disciplined next time. However, self-discipline guarantees the continuity of a daily routine. The Japanese are a very self-disciplined race: they have one of the lowest crime rates in the world and are well-known as 'obedient' tourists.

Photo 4.7 Good Example of Standardisation -- Department label (4.14)

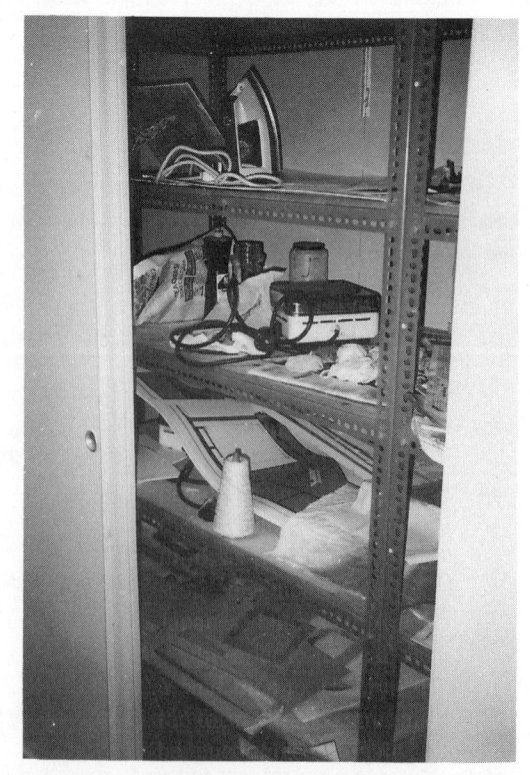

Photo 4.8 Bad Example of Standardisation -- Junk behind the door due to lack of transparency (4.1)

Chapter 4: Japanese 5-S Practice 77

Discipline is a process of repetition and practice. Think of discipline as an integral part of industrial safety. How many people have had accidents because they forgot to wear their safety helmet, their safety shoes, or their goggles? Far too many. How many have had accidents because they stuck their hands into the machinery without shutting it off first? Again, too many. It is important that everyone has the habit of obeying simple safety rules. An example of bad discipline is illustrated in Figure 4.3.

Figure 4.3 A Bad Example of Discipline -- No Self-discipline

McGregor [1960] identified that there are two sides of the human attitude towards work. In his Theory X, he observed that humans dislike work and would like to get away from work if possible. On the contrary, in his Theory Y, he observed that humans actually like working and they work as hard as they can to achieve results. This is the case when people are motivated to do their work. Ouchi [1981] observed many successful Japanese and American firms and found out that people actually consider the organisation as part of their family. The staff in these companies devote so much energy and time to their work that one might think as if it is their own business. This type of devotion to work represents the scope of Ouchi's Theory Z. His research shows that it applies not only to the Japanese but also to American workers.

In order to make a successful and painless transition from Theory X to Theory Y and then to Theory Z organisations should install some degree of discipline in the form of procedures and work instructions. Consequently, self-discipline should be encouraged. Finally, the employees will develop their own self-discipline framework.

Photo 4.9 Good Example of Discipline -- Seeing-is-believing (5.10)

Photo 4.10 Bad Example of Discipline -- Individual responsibility (5.7)

Daily 5-S Activity Checksheets

The checksheets in **Annex 4.2** (for Workshop) and **Annex 4.3** (for Office) at the end of this Chapter will facilitate introduction and development of self-discipline in everyday 5-S activities. Initially, it should be used by the supervisor as a checklist. The checksheet can be used by the staff themselves to remind them of their daily 5-S activities until most of the points are compiled regularly. In the next stage the staff will take a full 10-minute 5-S course and possibly modify and add-in some of the other 5-S activities which are relevant to their work.

The 5-S in the Office

The office is for administrative and managerial work, that has to be done effectively and efficiently. The accounting department, for example, has to do the monthly tallies clearly and accurately. They have to get the numbers out quickly and in a form that will give top management a clear picture of where the company is heading.

As a result, the office needs the 5-S to eliminate inefficiencies, to prevent mistakes, and to keep things running smoothly. The office is basically the same as the factory. They both get raw materials and turn out a product, the only difference being that the factory deals with tangibles while the office deals with information and paperwork. As such, the same 5-S principles can be applied to both factory and office.

The Nature of Office Work

There are things that complicate office work. Office work is harder to standardise, and it often appears inscrutable to outsiders. In addition, much of it is done by hand with wide discretionary approaches, and it is difficult for other people to help. Often there are no or only minimal operating procedures, working conditions vary widely, and the process is not mechanised. There are only few written standards. There is no foolproofing. All this considerably affects the initiation, acceptance and implementation of the 5-S practices in the office environment.

The office produces information which is usually used in-house rather than sold to outside customers. There may be some requirements regarding the source of information and timing, but there is very little monitoring along the way to ensure that everything goes smoothly and that no mistakes are being made. As a result, there is very often little awareness of the need for continuous improvement and the 5-S practice.

Main Points in Office 5-S Activities

There are a number of guidelines which can facilitate 5-S activities in the office environment. These are:
1. Reduce the number of ledgers, forms, tools, and the like to eliminate all the unnecessary clutter.
2. Pull out all the paperwork, files, tools, and the like and find better ways to store things. People should be able to get anything they need in 30 seconds. You should start a 'one-is-best' campaign to get people to write memos no longer than one page, to have office equipment kept in just one place, to standardise stationery, and to eliminate the idea that everyone has to have his own copy of everything.
3. Shift from the individual-based to group-based work. Standardisation and support system (e.g., multifunctional worker) are needed to even out the work load.
4. Organise and implement *continuous improvement* so that your ingenuity yields tangible results (the WHAT, WHY, WHO, WHEN, WHICH and HOW brainstorming would help). Make planning tables showing such things as what the plan is, how it is supposed to be implemented, and how much progress has been made. By breading the process down into steps, it should be possible to standardise and hence to create manuals detailing how each step is to be performed. There should also be a space to evaluate what is happening, since this will enable you to identify the trouble spots.
5. Study and improve office tools. Very few of the tools and ledgers that people use in the office were designed for that specific office. There is thus considerable room for customisation and improvement. Paperwork, ledgers, and tools should be created around the job, not the other way round. Many times, you may even find that the best way to improve a form is to eliminate it. Use your imagination, and aim for tangible results.
6. Try to keep a clean and orderly office. The office is the first place many visitors see. It shapes their first impressions. It sets the tone for the company. Yet that is just one reason why it is so important that you make a total effort to make sure that the office is clean -- meaning that there is no grime on the ceiling, the floor, the walls, the desks, the filing cabinets, or anywhere else and that everything is neat.

Even more crucial are the benefits that will accrue from neatness. Everybody will know where things are. Information will be accessible. And a neat office reinforces the pressure on people to keep things simple and to finish each task as it comes up rather than putting it on a 'look-at-later' mountain.

The One-is-Best Campaign for the Office

The one-is-best campaign is a means of streamlining operations and facilitating management. Basically, it is a way of avoiding redundancy in that it strips away

Chapter 4: Japanese 5-S Practice 81

the non-essentials and enforces essentials. There are all kinds of 'ones' involved here. Some of the most popular are used in the '5-S Audit Worksheet for Organisation'. They are:

- **One-set of stationery/tools**
  Limit people to, at most, one of each stationery and tools. For less-frequently used items, such as marker, strings, heavy-duty stapler, etc., one set can be shared among staff in an office.

- **One-location files**
  Centralise file storage in one location. This will cut down the number of files you think you need. Switch to good filing practices. Make your files immediately accessible. Limit distribution as this will also speed up recall if deem necessary. This is particularly important in computerised offices. You should avoid 'Islands of Automation' by making an effort to network computers so that the common files are centrally stored in the file server. Moreover, there need to be a systematic way of naming files so that everyone will be able to understand and locate files within 30 seconds. Obsolete files have to be purged, or only a sample file need to be kept for future modifications.

- **One-copy filing**
  Keep only the original of each document, not making a lot of photocopies. If at all feasible, do not print even the original, but just leave in the computer server (with backup copy on floppy). You will be surprised that even originals would have many copies of revisions.

- **One-page memos**
  Limit all reports, memos, and the like to one A4 size page. This will cut your paperwork dramatically. This can be achieved by improving the quality of your writing. Write from the reader's perspective and write reports that add value commercially. You must also limit the distribution to those who cannot do their work without having a copy. If the office is on E-mail, the best strategy is to create a paperless office by using the E-mail facilities as a matter of routine.

- **One-hour meetings**
  Limit meetings to one hour. This will cut the amount of time spent in meetings. Arrange procedures before the meeting. Follow up after the meeting. Reduce the amount of wasted time. Hold meaningful meetings. Review whether or not each meeting is really necessary. Limit participation. Sometimes, proper use of the telephone or E-mail facilities can be a good substitute for meetings and this should be considered first.

- **One-minute telephone calls**
  Try to gather your thoughts and questions before you pick up the telephone. Be concise and exact in your communication. Say good-bye within one minute if you can.

- **One-day processing**
  Institute same-day processing. This will dramatically reduce the amount of work held over. Reduce processing time. Reduce the number of sections that

have to see things and the number of people who have to sign on. Eliminate documentation grid-lock. Make the 'in', 'out' and 'pending' boxes smaller. If you are on E-mail, make sure that you check your E-mail letter box daily and reply immediately if at all possible.

Remember the office is management's home ground. If you cannot implement the 5-S in the office, how can you expect to do any better in the factory? Just as management has to take the initiative in instituting the 5-S company-wide, the 5-S effort has to start in the office.

Why is the 5-S practice useful?

There are many things that people do automatically without thinking. The 5-S can help in everything we do. The 5-S is like a mirror reflecting our attitudes and behavioural patterns. Even so, we all too often avert our eyes and prefer not to look at what we see there. Listed below are some features of a non-performing workplace:
- Bad interpersonal relations.
- People look worn-out.
- High absenteeism.
- Workers do not care about their work.
- Worker do not make suggestions on how to improve the work process.
- QCC activities are stalemated.
- The workplace is beset with defectives and reworking.

Firstly, the 5-S practice is useful because it will help everyone in the organisation to live a better life. If you implement the 5-S at your home first, you will see the real benefit. Once you gain experience and become aware of the usefulness of the 5-S, you can start implementing it at your workplace. In turn, you will benefit from the improvement in your workplace environment.

The 5-S can form a good basis for QCCs activities. Many a time circle members come together with a lack of agenda for discussion. The 5-S can provide a very good framework for improvement. This practice tackles the root of the problem, such as work organisation, neatness, cleaning, standardisation, and self-discipline.

The 5-S also forms a basis for other quality improvement activities. For example to implement ISO 9000, many consultants would advise their clients to hire a lorry before the compliance audit -- to carry away the rubbish. Obsolete documents and materials would always invite query from the auditors. Moreover, seeing-is-believing. If the firm being audited can project a neat and clean image to the auditors, there is little reason for them to doubt the effectiveness of the implementation of the ISO 9000 quality management system.

Practising the 5-S activities results in immense benefits to the company. A 5-S workplace is high in *Quality* and *Productivity,* keeps *Cost* down, ensures

Chapter 4: Japanese 5-S Practice

Delivery on time, is *Safe* for people to work, and is high in *Morale*. As Dr. Deming has pointed out: the management should try to develop an environment in which the workers *enjoy* their work and *take pride* in it. What is this environment, then? It is simply the 5-S environment! That is why Y. Kondo (1971 Deming award winner) said, "In Japan, top executives have cited the 5-S as their number one management priority."

The specific purposes of the 5-S practice can be related to the following aspects of any working environment: *safety, efficiency, quality and breakdowns,*

Safety

Safety is, in many respects, dependent on organisation and neatness of a workplace. What this really means is that you have to pay attention to the little things. Are you wearing your safety helmet, safety gloves and shoes? Are you being careful when you transport things? Are the paths clear? These seemingly insignificant things make the difference. That is why the Japanese are concerned with an orderly workplace. The 5-S is also important to personal safety and health by preventing electric shock, fires and slippage accidents due to poor insulation, oil leaks, pollution from filings and fumes, or anything else that threatens human health and safety.

Figure 4.4 Not following the 5-S rules could lead to hazard

Efficiency

The famous chef, the skilled carpenter, the great painter and the great writer maintain their tools properly. There are no rusted knives, no saws with teeth missing, no matted brushes, and no lousy computer. They use good tools, and they take good care of them. They do not have to waste time while they are working. They know that the time spent on looking after their tools is far less than the time wasted due to poor quality products and potential problems that can be created.

Quality

Modern electronics and other machinery demand very high levels of precision and cleanliness. Just a spot of grime can cause a computer to crash. Filings and burrs can mean that things do not fit tightly. Dropping faulty parts on an assembly line can mean that the wrong parts are put together or that the product is shipped to the wrong client. There are all kinds of major problems caused by seemingly minor 5-S lapses. It is clear that the 5-S is prerequisite to quality, and this cannot be overemphasised. "Good quality products come from good quality working environment".

Breakdowns

There is a common "Monday Morning Syndrome" at some manufacturing plants. This occurs where the machinery seems to stick, and where hydraulic and pneumatic equipment pressure levels are low on Monday morning. All of these things happen because the build-up of grime over the course of the work week has had time to harden and to settle into places where it should not be. All of these things happen because the company does not practise 5-S during the week.

People have the same problem. When they do something every day, they get into a routine. Yet when they come back after a vacation, they forget which way the valve should be turned, which way is off for the switch, and what readings are normal for the meter. All of these things should have been marked. But all too often, they are not. All too often, the workers assume that no labels are needed. And then when they come back from vacation, they find out how faulty memory can be.

How to Implement the 5-S

5-S implementation requires commitment from both the top management and everyone in the organisation. It is also important to have a 5-S Champion to lead the whole organisation towards 5-S implementation step-by-step. If you decide to be the 5-S Champion of your organisation, the following steps will help you to achieve success.

Step 1: Get Top Management Commitment and be Prepared

You have to sell the idea of the 5-S to the most senior executive of your organisation. Moreover, and like any other quality programme, it is no good to get just his lip-service. He needs to be 100% committed; not just in announcing

Chapter 4: Japanese 5-S Practice

the start of the 5-S practice in the promoting campaign, but committed to give resources for training and improvements. Then you need to get prepared yourself.

In promoting the 5-S activities, the important thing is to do them one at a time and to do each thoroughly. Even the little things have to be taken seriously if they are to make any meaningful impact. This process can be stratified as follows:

1. Make a decision and implement it (e.g., the decision to get rid of everything you do not need, the decision to have a major housecleaning, and the decision to have 5-minute clean-up periods).
2. Make tools and use them (e.g., special shelves and stands for things, instructional labels, and placement figures).
3. Do things that demand improvements as prerequisites (e.g., covers to prevent filings from scattering and measures to prevent leakage).
4. Do things that require help from other departments (e.g., fixing defective machinery, changing the layout, and preventing oil leakage).

The first two categories demand lots of participation. You need to involve everyone in these activities if they are to have any lasting impact. By contrast, the last two categories require lots of innovative ideas. You need to consider not only what has to be done, but when it might be done and how much it might cost. None of this is easy to decide on.. Yet each part of the process has to be facilitated. The clearly stated goals and defined ways of reaching them will help a lot. Decide which of these things you can do by yourself and for which ones you need help.

Step 2: Draw up a Promotional Campaign

The first thing to do for a promotion campaign is to set up a timetable. An example of a 5-S implementation plan is shown in **Annex 4.4**. In general, the plan can be broken down into 10 key activities:

1. Get top-management commitment, assess status quo and establish implementation plan.
2. 5-S Workshop for 5-S Facilitators -- based on the 5-S Audit Worksheet in Annex 4.1, identify the key 5-S activities, one from each of the 5-S for the first cycle of implementation.
3. 1st 5-S Day -- Organisation (e.g., Throw away things you do not need.)
4. Daily 5-S activities by everyone.
5. 2nd 5-S Day -- Neatness (e.g., Name everything and assign locations.)
6. 3rd 5-S Day -- Cleaning (e.g., All-together housecleaning)
7. 4th 5-S Day -- Standardisation (Visual management & transparency for things)
8. 5th 5-S Day -- Discipline (e.g., Do your own 5-S Audit)
9. Grand Prize Presentation for the best 5-S department/section
10. Review and plan for next 5-S Campaign

If your company operates on Saturday (half-day), this will be an ideal day to chose as a 5-S Day. Alternatively, you can use Wednesday afternoon if more suitable. However, it is important that everyone must do it. Those excused by their supervisor will have to do it on the next convenient day.

A week before the Day, posters should be put up to explain briefly the focus and activities of the Day and to give some simple examples. As for motivation, it is usually necessary to give prizes (gold, silver and bronze) to the top three 5-S winning teams of the Day. The judging panel should include the 5-S Champion and all the 5-S Facilitators. The prizes should be presented by the CEO at the end of the Day.

Step 3: Keeping Records

It is important to keep records not only of decisions made but also of the problems encountered, actions taken and results achieved. Only if past practice has been recorded people will have a sense of progress and improvement over time.

Photographs

Photographs are one excellent way of keeping records. There should be both full-view pictures of the shop and close-up pictures of specific parts and places indicating all the important phases of the 5-S implementation. These photographs not only provide points of reference for the people involved, but for outside experts as well. They can also be used to publicise the progress at company-wide 5-S meetings. Photo 4.1 to 4.10 illustrate the 5-S action in a company.

Videos

Nowadays video is used as a powerful tool to explain problems and convince an audience. Some companies find it rather encouraging and beneficial to record on a videotape the situation before the 5-S Day and during the course of action. Videos taken before the 5-S Day are effective in explaining some of the existing problems. Videos taken during the 5-S Day can illustrate the difference before and after the improvement. It will make a scene if used in the feedback presentation at the end of the Day. However, it is important that the background information and people involved are introduced, and the message is explained clearly during video recording. If necessary, the message can be supplemented by the presenter's explanation during replay. Experience proves that video recording and presentation adds much fun to the 5-S Day.

The 'P' Mark

Use a bright red or yellow **P** to indicate **P**roblem places and places that need attention. You might even have special days when this is done or special teams to do it, and the Ps should serve as vivid reminders of how much remains to be done. Conversely, a workplace with no Ps is either a workplace that has not been checked yet or a workplace that has already done as much as could be done -- and the difference will be immediately obvious to any observer.

Quantification
It is very important that you find ways to quantify what you are doing and the progress that you have made. This can be as simple as before-and-after comparisons to measure the amount of oil no longer lost to leakage or the amount of filings no longer being swept up. Use your imagination and be creative here in how you want to quantify these things to make the results more understandable. Apart from giving you a better insight into what is being done, numbers have a persuasiveness of their own that can be used to convince the sceptics.

Museum Rooms
As a last resort, you can keep some of the old tools and equipment in a special museum room -- a display of how primitive things used to be. This requires space, willingness and resources but such displays are more persuasive than photographs or statistics.

Step 4: 5-S Training

The 5-S activities are all directed at eliminating waste and effecting continuous improvement in the workplace. Right from the beginning there will seem to be lots of 5-S activities to be done. As you go on, you will notice that there are always additional 5-S problems to solve. They are not insurmountable, though, if considered and solved one at a time.

The 5-S activities are aimed at maintaining things at a desirable level. In promoting the 5-S activities, there is great emphasis on having people do things themselves and devising their own solutions. Yet this should not turn into a desire to discard anything that anyone else has made or done. Synergy of everyone's effort will result in a significant improvement of the organisation as a whole. True progress is achieved not in one great leap, but in small increments. Each subsequent step should be built upon the previous until you finally get where you want to be. And each step is more difficult than the one before.

It is essential in the 5-S activities that you train people to be able to devise and implement their own solutions. Progress that is not self-sustaining -- progress that always has to rely upon outside help -- is not real progress. It is important that your people know, for example, how to use the computer to do charts and graphs, even if it is not part of their job description. They need to study maintenance techniques. And oddly enough, the more problems they are capable of solving, the more problems they will spot.

Training should also include section-wide or company-wide meetings where people can announce their results. Not only does this provide incentive, but the exchange of ideas and information is often just what you need to keep everybody fresh.

Step 5: Evaluation

As with so many other things, it is very easy to get into a routine with 5-S activities -- particularly because they demand constant everyday attention to routine details. At the same time, because the individual tasks appear minor even though they have great cumulative impact, it is easy to think that you can put them off. Everybody is busy, and it is difficult to make alert 5-S activities a part of the daily routine. Workplace evaluations and other means are needed to keep everyone abreast of what is happening and to spot problems before they develop into major complications. In essence, you need to devise ways that will get everybody competing in a friendly but no less intense manner. Your evaluation tools are the key and it is as simple as using the 5-S Audit Worksheet as your evaluation criteria.

Patrols and Cross-evaluations
Two other techniques that you can adopt to promote the 5-S activities are patrols and cross-evaluations. Patrols can go around to the various workshops and offices and point out problems. This is similar to 'managing by walking around', but the patrol members do not even need to be management personnel. They simply need to know what to look for and have the authority to point out problems that need to be worked on. They simply need to know what questions to ask.

Cross-evaluations are a variation on this theme in that they involve having teams working on similar problems offering advice to other teams. One advantage of doing this is the exchange of ideas and mutual learning.

The objectives of the evaluation is to ensure that the 5-S implementation will lead to a conducive total quality environment (Figure 4.5).

Figure 4.5 Practising the 5-S leads to a total quality environment

Chapter 4: Japanese 5-S Practice

Three further Mini-Cases are given as illustration. Mini-Case 4.2 & 4.4 provide tips for 5-S implementation and Mini-Case 4.3 describes how 5-S was implemented at SIRIM and promoted at the national level.

■■

Mini-Case 4.2:

Even 5 Minutes Can Make a Difference   [Osada, 1991]

Organisation and neatness might not make much of a difference when you have all the time in the world, but it does count when you are working to a tight schedule. That is why it is useful to think of the 5-S as part of this schedule. Do not spend too much time on the 5-S activities. Set aside a short period of time when everyone will concentrate on the 5-S. One day it might be to make sure everything is in place. Another day it might be to make sure there are no oil leaks. It can be anything, but it is important that everybody be involved in this for a short time.

For example, some companies have instituted 5-minute 5-S periods. Everybody knows what they are supposed to do, and they know that they only have 5 minutes to do it. And they are all working on the 5-S at the same time. You will be surprised at how much can be accomplished if your people have had practice in the 5-S and they know what they are supposed to do. This is also a useful way to mark the beginning or end of the work day. It is only 5 minutes, but it means vastly better efficiency in the long run.

■■

■■

Mini-Case 4.3:

Implementing 5-S at the Standards and Industrial Research Institute of Malaysia (SIRIM)

This Mini-Case resulted from my experience in Malaysia. The SIRIM is the Government organisation in Malaysia responsible for the industrialisation programmes for Malaysian economy. It is one of the largest set-ups of this kind in the world, encompassing the functions of national standardisation, technology transfer and quality and productivity improvement consultancy. Apart from the Head-office in Shah Alam, SIRIM has seven branch offices throughout the country. At its head-office, there are over 30 blocks of buildings with nearly a thousand employees.

SIRIM has decided to implement the 5-S itself. This was necessary as SIRIM saw its benefits and also wanted to set example for industries. The training was a kick-off in a big way. The Director General of SIRIM announced a Saturday as the SIRIM 5-S Day. Before that day, I took some live videos of the 5-S audits conducted at the SIRIM head-quarters.

On the 5-S Day, the Director General took the lead and put on his sportswear. I gave a 40-minute seminar about 5-S to some 80 senior staff, including the 20-minute 5-S audit video presentation. At the same time, all remaining 900 staff were given a one page guideline to "throw away the rubbish and do all-together cleaning". At 1 p.m., when the 5-S Day was over, three lorry-loads of rubbish were thrown out of the 30 blocks of buildings. More interesting was that a week after the 5-S Day more rubbish came out of the buildings including steel file cabinets which were not suitable for the 'transparency' requirement of the 5-S.

Since the 5-S Day, many individual sections have requested in-depth workshops on each aspect of the 5-S practice. I started training a group of facilitators who has been actively carrying out the 5-S activities. There has been continuous requests from industries for consultancy service from the SIRIM's Industrial Extension Unit regarding implementation of the 5-S. As a result of the high demand for 5-S consultancy service, the Unit has been allocated additional human resources to satsify this need.

■■

■■

Mini-Case 4.4:

Implementing 5-S at the Wellex Corp. in the USA [Abramovitch, 1994]

Wellex Corporation was established in 1986 by three Taiwanese immigrants in the USA. By 1990, Wellex was an award-winning printed circuit board contractor with annual turnover of US$13.5 million and had built a good reputation among clients like IBM, Sun Microsystems and Silicon Graphics. As the business grew, the company employed 300 people from over 30 different countries.

In 1991, the demand for high-tech hardware plummeted. At that time, the only way to survive was to cut costs. Nevertheless, Wellex decided not to reduce staff since the management treated the employees as their biggest asset. In the attempt to cut costs, they turned to the Japanese experience.

In August 1991, the managing director and five managers visited Miyoshi Electronic, a Japanese company engaged in a similar business. There they were astonished by the cleanliness and neatness of the factory floor and the impact of overall organisation on employees' performance. The secret was in the 5-S system, the principles of which are basic for further quality improvement. The workers understood the importance of instilling the 5-S in their personal lives as

Chapter 4: Japanese 5-S Practice 91

well. It all contributed to improvement of interpersonal relationships across the company.

The team from Wellex was impressed and decided to launch the same practice back in their own company. Although there was a certain amount of scepticism towards the 5-S, a typical reaction of an individualistic culture, the results were encouraging. One of the assembly line workers explained the impact of the 5-S implementation as "Before, I'd have to wait around for my supervisor to tell me what to do. Now, I know what to do when I arrive in the morning. I have a schedule, and I keep records of all my work. This is a good system. Everybody knows what the problems are and how to fix them." They all emphasised the importance of organisation, cleanliness and discipline for a good atmosphere and mutual support in the factory. The 5-S system set different sets of rules from the previous practice but it made people more of a team. As another worker stated "Before the 5-S, we just worked. Now, I try to improve my work."

Just two years after the 5-S came to Wellex, productivity has skyrocketed by over 26%, with turnover exceeding US$23 million. This result proves that the 5-S culture is universal and can be related to any working environment if there is a commitment to the common objective.

■■

Chapter Summary:

▶ *5-S are organisation, neatness, cleaning, standardisation and discipline. 5-S is the key to total quality environment.*
▶ *5-S need to be assessed in detail. The 5-S Audit Worksheet is a basic framework for assessing the 5-S problems in your workplace.*
▶ *5-S in the offices are as important as those in the workshop and they need to be looked after. The one-is-best campaign is a very powerful tool to eliminate waste and ensure efficiency in the office environment.*
▶ *5-S is useful because it is the basis for other quality improvement programmes. The purposes of 5-S can be grouped under: safety, efficiency, quality and minimising breakdown.*
▶ *The five steps in implementing 5-S are: Get top management commitment, Draw up a promotional campaign, Keep records, 5-S training and Evaluation. The 5-S Implementation Plan is designed to do the first 5-S cycle in your organisation within 6 months.*

Hands-on Exercise:

1. Implement 5-S at your home first.
2. Then do 5-S at your own workplace.
3. Carry out a 5-S Audit in a department/section of your organisation.
4. Nominate yourself as the 5-S Champion of your organisation, follow the 5-S Implementation Plan and 'see-it-through' for the next 6 months.

Annex 4.1: 5-S Audit Worksheet

Firm Audited : _____

Address: _____

Phone No. : _____ **Date :** _____

Auditor : _____ **Audit No. :** _____

1. **ORGANISATION** (Seiri): Stratification management and dealing with the causes.

NO.	TYPICAL ACTIVITIES	LOCATION	AUDIT FINDING	Action by (& Date)
1.1	Throw away the things which are not needed.			
1.2	Deal with the causes of dirt leaks and noise.			
1.3	Organise cleaning the floors and housecleaning.			
1.4	Treat defects, leakage and breakage.			
1.5	Organise the storage of parts and files.			
1.6	One-is-best #1: one set of tools/stationery			
1.7	One-is-best #2: one page form/memo			
1.8	One-is-best #3: one day processing			
1.9	One-is-best #4: one stop service for customer			
1.10	One-is-best #5: one location file (e.g. LAN server for file sharing)			
Misc.				

Chapter 4: Japanese 5-S Practice

2. **NEATNESS** (Seiton): Functional storage and eliminating the need to look for things.

NO.	TYPICAL ACTIVITIES	LOCATION	AUDIT FINDING	Action by (& Date)
2.1	Everything has a clearly designated name & place			
2.2	30-second retrieval and storage			
2.3	Filing standards and control			
2.4	Zoning and placement marks			
2.5	Eliminate covers and locks			
2.6	First in, first out arrangement			
2.7	Neat notice boards (also remove obsolete notices)			
2.8	Easy-to-read notices (including zoning)			
2.9	Straight-line and right-angle layout			
2.10	Functional placement for materials, parts, tools, etc			

3. **CLEANING** (Seiso): Cleaning as inspection and degree of cleanliness.

NO.	TYPICAL ACTIVITIES	LOCATION	AUDIT FINDING	Action by (& Date)
3.1	Individual cleaning responsibility assigned			
3.2	Make cleaning and inspection easier			
3.3	Regular sparkling cleaning campaigns			
3.4	Cleaning inspections and correct minor problems			
3.5	Clean even the places most people do not notice			

4. **STANDARDISATION** (Seiketsu): Visual management and 5-S standardisation.

NO.	TYPICAL ACTIVITIES	LOCATION	AUDIT FINDING	Action by (& Date)
4.1	Transparency (e.g. glass covers for see-through)			
4.2	Inspection OK marks or labels			
4.3	Danger zones marked on meters and switches			
4.4	'Danger' warning signs and marks			
4.5	Fire extinguisher and 'Exit' signs			
4.6	Directional markings on pipes, gangways, etc.			
4.7	Open-and-shut directional labels on switches, etc.			
4.8	Colour-coded pipes			
4.9	Foolproofing (Poka-yoke) practices			
4.10	Responsibility labels			
4.11	Electrical/telephone wire management			
4.12	Colour coding -- paper, files, containers, etc.			
4.13	Prevent noise and vibration			
4.14	Department/office labels and name plates			
4.15	Park-like environment (garden office/factory)			
Misc.				

Chapter 4: Japanese 5-S Practice

5. <u>DISCIPLINE</u> (Shitsuke): Habit formation and a disciplined workplace.

NO.	TYPICAL ACTIVITIES	LOCATION	AUDIT FINDING	Action by (& Date)
5.1	All-together cleaning			
5.2	Do daily physical exercise all together			
5.3	Practise pick-up components and rubbish			
5.4	Wear your safety helmet/gloves/shoes/etc.			
5.5	Public-space 5-S management			
5.6	Practise dealing with emergencies			
5.7	Execute individual responsibility			
5.8	Good telephone and communication practices			
5.9	Design and follow the 5-S manual			
5.10	Seeing-is-believing: check for 5-S environment			
Misc.				

NOTE: Some of the above Typical Activities are not applicable to the office environment and therefore need not be audited for offices.

-- E N D --

Annex 4.2: Daily 5-S Activity for the Workshop

Name: _____ Dept./Section: _____

Week Commencing : _____ Checklist No.: _____

No.	Items to be Implemented	Mon	Tue	Wed	Thu	Fri	Sat
	5-minute 5-S						
A.1	Check your own clothing for condition and cleanliness.						
A.2	Check for any leakage or dropping, and pick up any parts, work, trash, or whatever that is on the floor.						
A.3	Wipe the meters, position markings, main places on the machinery and other important places with a rag.						
A.4	Wipe up any water, oil, or whatever that may have spilled or leaked.						
A.5	Realign anything that is out of place.						
A.6	Clean up the plates and labels and make sure they are clearly legible.						
A.7	Make sure all of the bits and tools are where they should be.						
A.8	Get rid of anything that is not needed there.						
	10-minute 5-S (on top of the 5-minute 5-S)						
B.1	Wipe the key parts and other places on the machinery with a rag.						
B.2	Fix any labels that might be coming off.						
B.3	Clean the floor.						
B.4	Get rid of the trash in the trash bins.						
B.5	Check the labels, instructions and oil inlets, and fix anything that is amiss.						

Chapter 4: Japanese 5-S Practice

Annex 4.3: Daily 5-S Activity for the Office

Name: _____   Dept./Section: _____

Week Commencing : _____   Checklist No.: _____

No.	Items to be Implemented	Mon	Tue	Wed	Thu	Fri	Sat
	5-minute 5-S						
A.1	Check your own clothing for condition and cleanliness.						
A.2	Check for any leakage or dropping, and pick up any parts, work, trash, or whatever that is on the floor.						
A.3	Tidy up and clean your desktop.						
A.4	Check where your documents and files are and put them back to their 'home'.						
A.5	Get rid of things you do not need, including personal effects in your desk drawers.						
A.6	Check where the filing cabinets, bookcases, and other furniture are and fix them if they are out of place.						
	10-minute 5-S (on top of the 5-minute 5-S)						
B.1	Get rid of anything that you do not need that is on top of filing cabinets, bookcases, and other furniture.						
B.2	Fix any labels that might be coming off.						
B.3	Clean the floor and your surrounding area.						
B.4	Check to see that the files are numbered, in proper order, and all accounted for.						
B.5	Check to see how many pencils, ball pens, erasers, and other things you have. Reduce them to 'one-is-best'.						

Annex 4.4: 5-S Implementation Plan

Company: _____ Dept./Section: _____

Issue No.: _____   Issued By: _____ Issue Date: _____

<======= MONTH =======>

Step	5-S ACTIVITY	Person(s) Responsible	1st	2nd	3rd	4th	5th	6th
1	Get top-management commitment, establish status quo and implementation plan.	CEO & 5-S Champion (5-SC)	•					
2	5-S Workshop for 5-S Facilitators	5-SC	•					
3	1st 5-S Day -- Organisation (e.g., Throw away things you do not need.) *	5-SC, Facilitators & CEO		•				
4	Daily 5-S activities by everyone	Facilitators		•	•	•	•	•
5	2nd 5-S Day -- Neatness (e.g., Name everything and assign locations.) *	5-SC, Facilitators & CEO			•			
6	3rd 5-S Day -- Cleaning (e.g., All-together housecleaning) *	5-SC, Facilitators & CEO				•		
7	4th 5-S Day -- Standardisation (Visual management & transparency for things) *	5-SC, Facilitators & CEO					•	
8	5th 5-S Day -- Discipline (e.g., Seeing-is-believing) *	5-SC, Facilitators & CEO						•
9	Grand Prize Presentation for best 5-S department/section	CEO & 5-SC						•
10	Review and plan for next 5-S Campaign	5-SC & Facilitators						•

* *Individual prizes (gold, silver and bronze) should be given to the top three 5-S winners for the Day. They should be presented by the CEO.*

Chapter 5

Business Process Re-engineering

Topics:

- What is BPR?
- Service industry and quality
- Best Practice Benchmarking (BPB).
- Total quality marketing
- Total quality purchasing
- Why is BPR useful?
- How to implement BPR

```
5-S
 ↓
BPR
 ┆
QCCs
 ↓
ISO
 ↓
TPM
 ↓
TQM
```

Introduction: What is BPR?

The definition of **Business process Re-engineering (BPR)** has already been given in Chapter 3. In simple terms, BPR is the radical redesign of business as a whole or individual work processes in order to maximise business effectiveness. It differs from continuous process improvement in that it challenges the need for the current process in the first place. If the process is indeed needed, BPR uses the sources of variation to produce Breakthroughs in process design. BPR's primary thrust is to alter processes, not to change an organisation's culture. However a climate that supports BPR should exist before such radical efforts have been attempted. Frequently, continuous improvement and BPR occurs separately. While each approach delivers benefits of its own, experience has shown that when they are done simultaneously the beneficial effects are magnified [Mutton, 1995].

Organisations must remain living, breathing, alert organisms if they are determined to handle what is going to happen. Nothing can be taken as permanent. A continual and formal preparation of the organisation to handle the future, along with re-evaluation of resource management, is the price of avoiding the quicksand. Rapid technological changes and competitive pressures of modern markets increase demands for quality, service cycle times and innovation. BPR is designed to help with these situations.

Unless organisations learn to identify and reform strategic business processes they will never be able to exploit changes in the market place. Profitability, market share and goodwill are not bestowed on organisations by some divine right. Instead the market awards these benefits to organisations that listen to and satisfy the voice of the customer. Change will ultimately affect every organisation. The ability to change and adapt rapidly is fast becoming a basic survival skill. Hence the importance of combining TQM and BPR.

To systematically enhance business processes today, there are some strategically important questions to be asked before initiating process improvement efforts. They cover three areas, as follows:

1. **The nature of the business.**
- What business are we in?
- How fast is our business changing and why?
- What are our strengths and weakness?

2. **The critical success factors**
- What are the key success factors customers are looking for?
- What are our current customers asking of us and other suppliers?
- Will they be asking the same in the future?

3. **The processes to be improved**
- Where could the competition hurt us?
- How can we effectively apply continuous process improvement?
- Which processes need re-engineering?

Figure 5.1 Re-engineer your business to the needs of your customers
[Lip, 1989]

Chapter 5: Business Process Re-engineering (BPR)

Continuous improvement of business processes is an organic part of total quality by definition. The answers to the questions above will assist in determining if and where process re-engineering should appropriately replace continuous improvement. Normally we would re-engineer if:
- The process is strategic.
- There is a high correlation between customer defined key success factors and the company's weaknesses.
- The process has a high price of nonconformance attached to it.
- The process may soon become strategic due to change.

If BPR is being carefully incorporated into the company-wide TQM process, as the TQMEX model suggests, organisations will ensure the effectiveness of one of the most potent weapons available in the corporate armoury to meet the challenges of the future. Figure 5.1 illustrates this point.

BPR or Continuous Improvement Dilemma

It has been already mention in the previous section that improvement is a key term in quality thinking. Under traditional thinking the internally focused workforce learnt to work to a standard which inevitably became a ceiling for performance. To shatter this ceiling, management must recognise that the issue of quality rests upon understanding what you do and working upon improving it.

However, improvement cannot happen where the culture has not altered. The core issue is, again, management thinking. As quality rests on the value of the service delivered to the customer BPR consists of identifying how well service performance is being achieved. Change happens quickly in those organisations that listen to their customers and take action from what they learn. The basis of their successful culture change lies in their understanding that customer satisfaction data drives action. Key issues can be identified through the accumulation of service performance data, and these data can be used as an itinerary for instigating change. In this way, continuous improvement -- *kaizen* becomes an integral part of the business operations.

One measure of the dynamics of change is by the Velocity Ratio which is defined as:

$$\text{Velocity Ratio} = \frac{\text{Value Added Time}}{\text{Total Elapsed Time}}$$

Velocity Ratio is a critical performance measure, estimated to be around 0.05 to 0.07 for many manufacturers [Wong, 1995]. In other words, only 5--7% of time is actually spent doing things the customer is paying for. In many companies, there is a focus for improvement to reduce waste and increase the velocity ratio to near the ultimate of **1.0**. Both BPR and kaizen techniques can help. BPR looks at the basics of the problem and tackles it in a radical way, while kaizen leads to the achievement of the objective more slowly in many small step changes. A comparison between the two improvement processes is shown in Figure 5.2.

		Continuous Improvement (Kaizen)	BPR
Common Point		Focus on the customer	
Starting Point		Not under competitive pressure to take immediate action	The company has the resources to handle the transformation
D I F F E R E N C E S	Strategy	Continuous small steps	Infrequent big leaps
	Approach	Start with what you have	Start with 'clean sheet'
	Methodology	Change what you have and learn	Forget and start again
	Process	Simultaneous processes	Selective, one at a time
	Value Added	Eliminate non-value added processes	Minimise inputs, add value to outputs
	Human Resource	People involved in the operations	BPR project team
	Technology	Less technology required	More technology required

Figure 5.2 Kaizen vs. BPR -- Similarity and Differences

Figure 5.3 BPR vs. Kaizen -- If you are nearly drown, kaizen won't help, you need BPR

Kaizen is a right thing to apply, if you are the world leader. It is still feasible if you are lagging, provided that you are prepared for extra efforts. You should consider BPR, though, if you are far behind the world standard in some or all of your processes (see Figure 5.3). In this case, it is the role of top management to identify those business processes that should undergo re-engineering, and then steer the company towards that goal.

Improving Customer Satisfaction through BPR

BPR challenges managers to rethink their traditional methods of doing work and commit to a customer-focused process. Many outstanding organisations have achieved and maintained their leadership through BPR [see Oakland, 1995]. Companies using BPR have reported significant bottom-line results, including better customer relations, reductions in lead time to launch a new product, increased productivity, fewer defects/errors and increased profitability. BPR uses recognised techniques (explained in the rest of this Chapter) for improving business results and questions the effectiveness of the traditional organisational structure. Defining, measuring, analysing and re-engineering work processes to improve customer satisfaction pays off in many different ways.

BPR starts off with a well understood Vision which should induce a clearly defined Corporate Mission Statement (which can become the Quality Policy Statement as in the case of ISO 9000 quality management system).

BPR breaks down the internal barriers that arise from line organisation and encourages the entire organisation to work as a cross-functional team with a common view of the business. It requires shifting the work focus from managing functions to managing processes. Process owners, accountable for the success of major cross-functional processes, are charged with ensuring that employees understand how their individual work processes affect customer satisfaction. The interdependence between one group's work and the next becomes quickly apparent when all understand who the customer is and the value they add to satisfy the customer.

Value Engineering

BPR aims at increasing the **Value** of the products and services to the customers. **Value Engineering (VE)** is a way to help achieving this goal. Value is defined as the *ratio of the output (functions) of a process to its input (cost)* and VE is defined as *the process to improve the value of goods and services.* Because of this definition, VE is also called **Value Analysis** or **Value Added Analysis**.

By definition,

$$\text{Value} = \frac{\text{Functions}}{\text{Costs}}$$

This is a simple but very useful relationship between the functions of a product and its cost. There are five ways to <u>increase</u> Value by VE.
1. **Reduce functions and reduce cost more** -- This is a "cut corner" exercise. Unfortunately, in today's competitive market, consumers are more demanding than before. So this is not an attractive solution and would reduce the image of the firm.
2. **Keep functions unchanged and reduce cost** -- This is typically a cost reduction exercise and should be implemented without the consideration of VE.
3. **Increase functions and reduce cost** -- This is, in most cases, impossible to achieve, unless the original design is bad.
4. **Increase functions and keep cost unchanged** -- This is difficult to achieve unless there are some dramatic design improvements.
5. **Increase functions more than cost increase** -- Most Japanese firms have a firm belief in this approach because in today's buyers' market, every consumer wants the best value for money and would like to increase the functions of their purchases. Consumer demand is further supported by intense competition in the market place. This is obvious by looking at the full range of electronic consumer goods, particularly in audio-visual equipment. For instance the camcorders today are smaller, have multi-function, professional quality and resolution, yet with little increases in price. There are numerous examples of this kind from Japanese products.

From the above analysis, the *5th option* is the most promising strategy in most cases. In Taiwan, a electric fan manufacturer adopted the same strategy in their product development in 1985 [Yip, 1986]. They put infra-red control in their free-standing fans: so that the customer can control the speed, the angle of swing, and even the decorative lighting at his seat. It may be considered as a gimmick to some people. However, the end result is that the company managed to sell 200,000 sets to the Japanese market which is highly competitive. Very soon, all major Japanese manufacturers have to put remote control onto their high-end fans! This example should give light to companies from other countries on how they can market their products to Japan.

Service Industry and Quality

Quality has traditionally been related to the manufacturing industry. Due to the improvement in technology and automation over the last century, more and more of the workforce in both the developing and developed nations have shifted towards the service industry. This trend will continue. It is therefore important to understand the special characteristic of quality as applied to the service sector.

Service is a major sector in today's world economy. For instance, in the USA, some 60 million people work in the service industry with about US$3 trillion of the Gross National Product in the private sector. US Government adds another

Chapter 5: Business Process Re-engineering (BPR) 105

30% to the employment. [Vroman, 1994]. The more complex the international economy, the more services are needed to address problem areas. Law, health, accounting, and education are some of the professional service areas that parallel retail stores, financial institutions, and charitable organisations.

Figure 5.4 illustrates the difference between manufacturing and service organisations. One primary difference is that products are tangible, and services intangible. Another difference is that the manufacturing system is relatively closed, and the service system relatively open.

MANUFACTURING INDUSTRY	SERVICE INDUSTRY
Products are transformed from materials.	Services do not exist until they are provided at the call of the customer.
They have physical dimensions and attributes, take up space in inventory, are depreciated, and often wear out.	They take up no space, can not be inventoried, and have no shelf life.
Products can be evaluated against specifications and criteria.	Service quality is evaluated against satisfaction of the customer.

Figure 5.4 Comparing Manufacturing and Service Characteristics

The challenge of business is to deal effectively with customers. Customer standards are usually vague, transitory, and changing. Identifying and controlling the factors that govern standards is very difficult. The challenge of service is translating customer needs such as timeliness, cleanliness, and friendliness into service standards.

Both manufacturing products and services are sold on benefits. Service is categorised as a variation of manufacturing. Many service organisations used the manufacturing model approach and profited by it. The fast food store epitomises the service use of the manufacturing model. Make the hamburger efficiently, with low skill and minimum pay, and then market it heavily. Fast food outlets like McDonald's and Burger King concentrate on manufacturing meals [Ho, 1993]. Marks and Spencer used a similar philosophy in managing its stores during the 1980s and is widely known as the 'factoryless manufacturer'.

The competitive advantage in the service organisation lies in its quality leadership in the market place. How the service is delivered by the front-line person is the major issue. For instance, all of the above three companies are fully aware of this trend and have intensive training on their staff to provide quality service delivery. Every MacDonald's employee knows very well that their corporate mission encompasses Quality, Service, Cleanliness, Value-for-money (QSCV). *The most important thing to know about intangible products is that the customers usually do not know what they are getting until they don't get it.*

Value in Service Industry

Like its manufacturing counterparts, more successful service organisations have adopted a new model since the 1980's. They shifted the emphasis from financially optimising assets to customer satisfaction. The emphasis has shifted from emulating old-design manufacturing to making the core technology flexible, highlighting the key role of the front-line employee to satisfy customers.

Service standards are more demanding because customers are becoming increasingly sensitive to quality. The focus has shifted from efficiently "manufactured" service items towards a more flexible model of responding to a more quality sensitive customer. At Rover Car, they use the term *"Extraordinary Customer Satisfaction"* to express its focus on the customer. The system has moved from the closed-system of the old-design organisation to the open-system of the new design, incorporating the customers' requirements at the centre of their operations.

Quality processes are directly related to value. The value-added approach can be either explicit or implicit. Explicit value can be measured and can be clearly defined or visible. Implicit value is not easily converted to currency and is related to perceived performance needs or image. At consumer level, there have been growing indications of dissatisfaction with the treatment of customers and clients in banking, transportation, and health-care services as well as in other sectors. (For details see *Business Week*, 11 Nov 1991.)

QUALITY POINTS	SERVICE ORGANISATIONS
Design it in	Heavy emphasis in service re-design on meeting changing needs and constant revision of service attributes. *Provide low-fat food to health-conscious customers and monitor changes in taste.*
Build it in	This refers to the service delivery. The customer is part of this setting. The front-line person is a major player. *A dinner is cooked and served according to the customer requirements.*
Inspect it in	Inspection occurs during the delivery of the service. This function is minimised in the new-design setting. *The responsibility for quality rests with the cook and the waiter.*
Fix it in	In the new-design organisation, 100% satisfaction is guaranteed. *If not satisfactory, the meal must be immediately replaced to satisfy the customer.*
Educate it in	Inform the customers about the changes and the benefits. *Introduce the new food range (see the first point above).*

Figure 5.5 Building Quality into Services

Figure 5.5 lists the five ways in which an organisation, such as a restaurant, can incorporate quality into its service. Strategic resource allocation in the organisation determined to meet/exceed customer needs will enable a careful design and flawless delivering of the service. Such a strategy limits, if not eliminates the need for resource allocation to inspection and fixing activities.

Many organisations work hard to design quality in and think through delivery barriers. For example, the Ritz Hotel builds quality into its operations by allocating major resources to service recovery. It recognises that its customers will not come back if Ritz makes errors or creates an uncomfortable atmosphere. Because of this approach, every front-line employee can cancel up to £1000 of a guest's bill on their own authority to solve a problem, with no questions asked. This is equivalent to the unconditional guarantee of some manufacturing firms. At Safeway Superstores, they guarantee *'Refund & Replace'*. If their customer finds some defects in the goods he has purchased, he will get an appropriate substitute plus a full refund. A more detailed example of satisfying customers is shown in Mini-Case 5.1.

■■■

Mini-Case 5.1:

Service Quality at Scandinavian Air System [Carlzon, 1987]

In the airline business, one seat on an aeroplane is much like any other. It is passenger service quality that makes the big difference in the treatment by front-line personnel. This includes situations on board, at the ticket counter, and at the baggage area. The international airline's challenge is complex. What has to be done to ensure the traveller's complete satisfaction with his long hours' experience of flying 10,000m above the sea level at 800km/h.

*In Scandinavian Air System (SAS), each time an employee interacts with a customer is noted as a **'moment of truth'**. Jan Carlzon, President of SAS estimated that some 50,000 moments of truth occur daily as SAS personnel interact face to face with customers. Carlzon said, SAS is not the aeroplane, or the airport gate, or the overhaul station. It is the contact between the employee and the passenger. The moments of truth occur when the customer, consciously or unconsciously, has a need and turns to an employee or the physical facilities for a solution. The moment of truth is successful if either the design of the physical facilities anticipated those needs, or the service plan provides anticipatory human intervention.*

■■■

Customer Satisfaction and Retention

Customers expect certain things and if the organisation does not meet those expectations, the customers are dissatisfied. The only way for the organisation to keep a high level of customer satisfaction and still operate efficiently, is in mastering the art of an optimum level performance that ensures that customers' expectations are consistently met. Parasuraman et. al. [1990] describe the problem as "the extent of discrepancy between customers' expectations or desires, and their perceptions of the quality of the service". The means by which customer expectations are generated include:
- word-of-mouth communication,
- personal needs,
- experience, and
- external communications that influence customers' expectations.

Friends, consumer groups, and the government play a role in shaping expectations. Customers will shop in places where service standards are designed to meet such expectations. High discrepancy between expectations and perceptions of the service results in customers dissatisfaction. Le Boeuf [1987] cited a study on the reasons why customers quit the relationship with a particular supplier:
- 3% move away giving no reasons,
- 5% develop other supplier relationships,
- 9% leave for competitive reasons,
- 14% are dissatisfied with the product,
- 68% quit because of an attitude of indifference toward the customer by the owner, manager, or some employees.

Le Boeuf claimed that businesses spend six times more to get new customers than to keep old customers. Customer loyalty is worth ten times the price of a single purchase, in the average case. There is an old saying in service industries that if customers like the service, they will tell three people. If they don't like the service, they will tell eleven people. It illustrates the different "speed rates" of spreading around good and bad news about the company.

There is an opportunity for service recovery, if the firm is listening to the customers. Some research findings reveal that about half of the dissatisfied customers just go away. Disneyland is effective in bringing people back to its parks. Red Pope [1979] says "When a company's customers are happy with the service and product, and find enthusiastic and knowledgeable personnel who are anxious to help, chances are that company will continue to enjoy the lucrative patronage of those customers for a long, long time."

The SERVQUAL Model

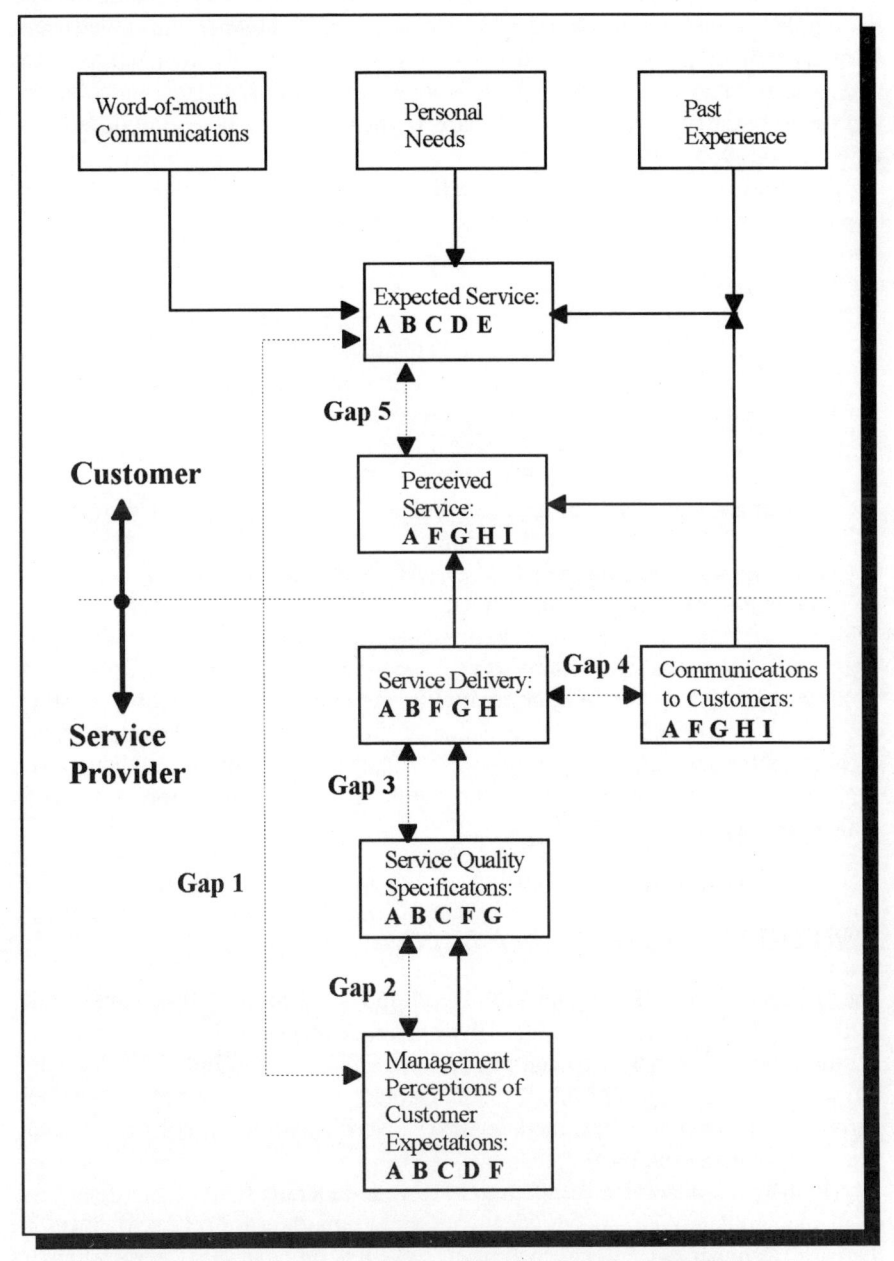

Figure 5.6 Conceptual Model of Service Quality, **SERVQUAL**

In the TQMEX model, BPR needs to establish a set of customer oriented business objectives. In the service sector, such objectives can be derived from the service specification. Service specification forms an integral part of the Service Quality (SERVQUAL) conceptual model (Figure 5.6). This model measures tangible and intangible service elements (Parasuraman, *et al.* [1990]). It investigates discrepancies or **gaps** in the consumer-supplier chain to highlight target areas where quality may be improved. The model shows that the customer has expected service **A-B-C-D-E**, but at the end, as a result of the 5-gaps in the operations, received **A-F-G-H-I** instead!

Parasuraman *et al.* also suggest that the criteria used by consumers which are important in moulding their expectations and perceptions of service fit **ten dimensions:**
1. Tangibles: physical evidence.
2. Reliability: getting it right first time, honouring promises.
3. Responsiveness: willingness, readiness to provide service.
4. Communication: keeping customers informed in a language they can understand.
5. Creditability: honesty, trustworthiness.
6. Security: physical, financial and confidentiality.
7. Competence: possession of required skills and knowledge of all employees.
8. Courtesy: politeness, respect, friendliness.
9. Understanding/knowing the customer, his needs and requirements.
10. Access: ease of approach and contact.

These ten dimensions vary with respect to how easy (or difficult) it is to evaluate them. Some, such as tangibles or credibility, are known in advance, but most are experience criteria and can only be evaluated during or after consumption. Some such as competence and security may be difficult or impossible to evaluate, even after purchase. In general, customers rely on experience properties when evaluating services.

SERVQUAL Applications in Education

The gap analysis for higher education based on consideration of **five gaps** is now used as an illustrative example [Ho, 1995]. Higher educational institutions have a multiplicity of customers' expectations and perceptions to consider. Teaching and learning plans for each individual module would specify the aims and objectives, purpose, credit value and module level, weekly syllabus, delivery format, assessment format and reading list.

The benefits of service specification are wide-reaching, and of significant use to all stakeholders. The stakeholders concept is important for service quality. In the case of higher educational institutions taken as an example, the stakeholders are: students, parents, sponsoring employers, employers of graduates, government bodies, franchise colleges, exchange colleges, academic staff, management, administration, professional bodies and other educational institutions.

Chapter 5: Business Process Re-engineering (BPR)

In accordance with the SERVQUAL Model, the five gaps are:

Gap 1 : Between customers' expectations and management's perceptions of customers' expectations. An example of this gap is on courses requiring hands-on experience, such as computing and design, where students may, erroneously, expect a greater emphasis on practical and technical study at the expense of academic rigour.

Gap 2 : Between management's perceptions of customers expectations and service quality specifications. Here, both students and future employers are the customers of the educational system. Students' lack of understanding and employers' particular focus and requirements may bias or limit expectations of both. As experienced professionals, academic staff should play a major role in designing and determining the format of the modules and courses which they deliver to narrow the expectations -- specification gap. In some instances, such a gap is desirable as both students and employers (customers) need fresh information inputs and knowledge from the educational system (supplier).

Gap 3 : Between service quality specifications and service delivery. In a modular system, it is not uncommon for a module designed by one academic to be taught by another member of staff who may apply their personal interpretation or focus of expertise to delivery of the specification. Continuous improvement of module should be encouraged, with the specification regularly reviewed.

Gap 4 : Between service delivery and external communications to customers. In order to recognise and minimise the opportunity of any such gap occurring, two-way informal and formal feedback systems are necessary.

Gap 5 : Between customers' expectations and perceived service. For instance, students who use laboratory, studio and workshop facilities in periods of peak demand may perceive that facilities are inadequately resourced whereas, in reality, facilities might be operating at, say, 50% capacity over the working week.

Almost all regular activities in the academic cycle may be analysed to determine which operations can benefit from improvement. Merely closing the gaps described above and satisfying specifications does not constitute a quality service. Quality is delighting the customer by continuously meeting and *improving* upon agreed requirements. Hence, flexibility should be inbuilt to exceed specifications and to respond to changes in service specification. Where a financial cost is attached to additional performance, such as an additional fee, this must be tempered against the scale of additional expenses and the willingness of the customer to meet this cost. Another example of SERVQUAL applied in education is shown in Mini-Case 5.2.

■■

Mini-Case 5.2:

The School of Quality   [Bonstingl, 1995]

American educators are beginning to realise that current systems of instruction do not encourage, or in some cases even permit, quality education. Many of them are now examining TQM as a possibly workable philosophy to create a new type of schools -- Schools of Quality -- which are based upon a way of schooling that better suits the imperatives of the 21st century than the factory-model system of schooling currently practised in many parts of the world.

What are such 'Schools of Quality' like? How do they differ from today's schools? Schools of Quality are grounded in four fundamental assumptions which Bonstingl called Four Pillars of Schools of Quality.

*The first Pillar of Schools of Quality is a **customer-supplier focus.** The entire school organisation must be dedicated to meeting human needs by building relationships of mutual support with people inside and outside the school. The role of a student is dual. As a worker, the student's product is his personal growth and continuous improvement that he presents to teachers or future employers (customers). As a customer he expects high quality teaching, security, and care from the school staff (workers/suppliers). Satisfaction of customers in both cases is required. This may not be fully symbolised by letter or number grades. Marks, in fact, detract from the prime objective of education which is developing young people's sense of taking pride in successfully completed job. In Schools of Quality, tests and other assessments are much more an indication of the teacher's success through the success of their students.*

*The second Pillar is a personal dedication by everyone to **continuous improvement**, little by little, day by day, within one's sphere of influence. For instance, student groups and teacher groups might form Support Teams to provide mutual support in academic and personal matters on a regular basis. Continuous improvement is much easier to integrate into operations if people interact and share experience.*

*The third Pillar is a **process/system approach.** Deming has hypothesised that as much as 80% of all things that go wrong in any organisation are not entirely attributable to individuals, but rather to the system in which they work. In schools, teachers and students combine efforts to continuously improve the system of teaching and learning, as teachers and administrators work together to improve the system of rules, expectations, policies, and other factors which constitute the operational culture of the school. Parents, families, business leaders, and the people of the community are invited to join this collaborative work for the long-term benefit of the young people and generations to follow.*

*The final Pillar is **consistent quality leadership.** This is the most crucial of the four Pillars. The ultimate success of the ongoing quality transformation is the* *responsibility of top management, and can only be achieved over time through constant dedication to the principles and practices of TQM. Leaders must construct fearless work environments in which coercion is set aside to permit risk-taking and temporary failures leading to continuous improvement. Consequently, school management should encourage the design of innovative curriculum, the use of new teaching methods, and more joint projects with external organisations and potential employers of the students.*

A decade ago, Americans were shocked by a report on the condition of education. The report was titled "A Nation At Risk" [DoE, 1982]. Today, educators have the opportunity to combine efforts with one another, with students and their families, with businesses and community people, and with others whose common future depends to some extent upon what is done in our schools. We have the opportunity today to transform our Nation At Risk into a Nation of Quality, beginning with Schools of Quality that are grounded in the philosophy of TQM.

■■

Service quality is important in contemporary organisations because customer expectations are demanding and competition is increasing. However, this is part of modern day business environment. Organisations should see this trend as opportunity rather than threat. This can be illustrated by the following story:

> *Once upon a time, an American and a Japanese were exploring in an African jungle. Suddenly, they saw a fierce, hungry lion. Having realised what was going to happen, the Japanese put on his running shoes immediately. The American laughed at him and said, "Do you think you can run faster than the lion?" The Japanese answered, "I am all right, as long as I can run faster than you!"*

Although this section is focused in service quality, the idea is equally applicable to the manufacturing industry. After all, manufacturing firms always have an element of service to satisfy their customers. Moreover, there are many manufacturers who provide services as well, such as after-sales service and design service, etc. Finally, service quality determines the customer requirement which is an integral part of the TQMEX Model.

Best Practice Benchmarking

Do you want your company to be a world beater? Of course you do. But where do you start? Or, to put it another way, how do you know whether your company now is good, bad, indifferent or at the top of the league in the things that matter in your sector of industry?

The answer to both questions lies in Best Practice Benchmarking (**BPB**). BPB is a technique used by successful companies around the world -- in all sectors of business -- to help them become as good or better than the best in the world in the most important aspects of their operations. Many of these companies are multi-national giants, such as Ford, ICI, Nissan or Xerox. Others are small businesses employing a handful of people. They all have one thing in common: profitability and growth come from a clear understanding of how the business is doing, not just in comparison with the previous year's performance, but against the best they can measure, too.

How do they do it? BPB comes in many forms, but essentially it will always involve:
- Establishing what makes the difference in their customers' eyes between an ordinary supplier and an excellent supplier
- Setting standards in each of the things according to the best practice they can find
- Finding out how the best companies meet these challenging standards
- Applying both other people's experience and their own ideas to meet the new standards -- and, if possible, to exceed them.

Companies world-wide have found that BPB offers:
- Better understanding of their customers and their competitors
- Fewer complaints and more satisfied customers
- Reduction in waste, quality problems and reworking
- Faster awareness of important innovations and how they can be applied profitably
- A stronger reputation within their markets, and
- As a result of all these, increased profits and sales turnover.

Making BPB Work

There are five key steps to the process of BPB. Each is relatively straight-forward and applied to both large and small companies. Large companies will tend to gather greater quantities of information and will be more concerned with issues of competition. Small companies will tend to focus on a few critical areas and be more concerned with operational improvements. The five steps of BPB are:

Step 1: Select what to benchmark
To work out exactly where to focus your efforts, ask these two key questions:

Chapter 5: Business Process Re-engineering (BPR) 115

- What would make the most significant improvements in your relationship with customers?
- What would make the most significant improvements in your bottom line?

Benchmarks important for customer satisfaction might include:
- Consistency of product,
- Correct invoices,
- On time delivery,
- Frequency of delivery, and
- Speed of service.

Benchmarks important for direct impact on your cost might include:
- Waste and rejects,
- Inventory levels,
- Work in progress,
- Cost of sales, and
- Sales per employee.

Step 2: List who to benchmark against

The following are some important considerations in choosing companies with whom to compare yourself:
- Do they know us? (Customers and suppliers should already have a good relationship with you.)
- Is their experience really relevant?
- Are they still at the activity you want to measure? A company can live off its reputation for a long time.
- Are we legally able to exchange this kind of information with the company?

In practice, there are four different types of an organisation you can benchmark against:
- Other parts of your company. For example, Rank Xerox radically improved time taken to develop its new products by benchmarking against Xerox's Japanese subsidiary, Fuji Xerox.
- Competitors. While for large companies this is often the most useful form of comparison, for small companies it may be least useful. In the free market competition, small companies should base benchmarking on the customers' standards instead.
- Parallel industries. Many of the best ideas for radical improvements come from this source. While companies in your own industry will generally tackle the same problems in the same way, companies in parallel industries may have very different approaches. For example, BAA, which administers seven British airports benchmarks itself against Wembley Stadium.
- Totally different industries. Here you will want to compare against very specific activities. Motorola, for example, exchanges information with other companies on the time between customer order and delivery. Among them are some of the best manufacturers of fast moving consumer goods, such as food supplied to supermarkets.

Step 3: Get the information

A great deal of the information you need is already there for the asking, in magazines and newspapers, in trade association reports and in specialist databases. You can also gather information from customers, suppliers and other observers. Your trade association librarian can often help to identify where to look; so can broader resource centres, such as the Institute of Marketing's Information Centre and the CD-ROM search facility available in most academic libraries. Benchmarking involves a variety of sources, constantly double-checked against each other. One source will reinforce another. Before investing time in a company visit it is essential to have conducted extensive desk research to know what to look for and the right questions to ask.

Step 4: Analyse the information

Too much information is as bad as too little. Gather only the information you need to make a direct comparison of performance. Once you have identified that other companies have a higher level of performance in a particular activity, you should:
- Quantify it, as closely as possible,
- Make sure you are comparing like with like,
- Get other managers' opinions on the lessons you can learn from the other companies' experience. The key question here is: 'how much of this is genuinely applicable to us?'

Step 5: Use the information

If you treat the process as a challenge rather than an indictment you will find that most people respond enthusiastically. So you have to make sure that you:
- Set new standards for the performance you expect and communicate them to everyone concerned in your organisation, along with an explanation of why standards have been raised,
- Make someone in authority responsible for devising an action plan to reach the new standards
- Provide the resources for employees to carry out additional research, if needed,
- Monitor progress so that the plan is implemented effectively.

You must continue to benchmark, to ensure that you can react to and keep up with other people's improvements. At the same time, you will need to keep challenging your own best practice to see if it can be improved upon through ideas generated by your employees. Being the best is tough; either you keep going up or you quickly start to slide down. Rank Xerox has established a pioneer model of BPB as shown in Mini-Case 5.3.

Mini-Case 5.3:

Rank Xerox's BPB   [DTI, 1994, Nov.]

Rank Xerox is part of Xerox Corp., a multinational company that found itself in deep trouble in the late-1970s. From the mid-1960s to the mid-1970s its profits rose 20% a year, not least because it had a near-monopoly on photocopier technology. By 1980 it saw its market share halve, as aggressive competitors moved in and beat it on price, quality and other important measures.

Xerox's solution was to benchmark the way its photocopiers were built, the cost of each stage of production, the cost of selling, the quality of the servicing it offered, and many other aspects of its business against its competitors and against anyone else from whom it could learn. Whenever it found something that someone else did better, it insisted that the level of performance became the new base standard in its own operations.

BPB has now become an every day activity for every department in Xerox and Rank Xerox. The guiding principle is: 'Anything anyone else can do better, we should aim to do at least equally well'. It is closely tied into the company's quality management programme, because BPB is one of the most important ways of identifying where quality improvements are needed. Not only has Xerox world-wide improved its financial position and stabilised its market share, but it has increased customer satisfaction by 40% in the past four years.

Typical of the way Rank Xerox uses BPB is a recent study of distribution, as explained by John Welch, Quality Manager. "We compared our distribution against 3M in Dusseldorf, Ford in Cologne, Sainsbury's regional depot in Hertfordshire, Volvo's parts distribution warehouse in Gothenburg and IBM's International warehouse and French warehouse. We found, for example, by comparing with the best that:

- We had an extra stocking echelon, which could be removed (i.e., we had international, national, regional; they had only international and regional.)
- We took one extra day in information flow between the field and centre, so we need to update our systems.
- They had transport logistics as a board level function.
- Warehouses became efficient not through a high level of automation but through efficient manual routines.
- 'First Pick' availability of parts averaged 90% in the best warehouses; we made only 83%.

Now we are putting those lessons to work, in upgrading our operations to be at least as good."

Total Quality Marketing

TQM is about applying Total Quality (**TQ**) concepts to all aspects of management. This would include all the business functions, such as marketing, design, research and development, production/operations, finance, human resources, administration, etc. However, since TQM is also about satisfying the needs of customers, the search for the real customer needs is of first priority. By definition, marketing function (but not exclusively marketing department) is charged with the same objective, to identify the needs of the customers. Therefore, identifying the needs must be seen as the first step towards BPR and should not be perceived as a responsibility of the marketing department. *Everyone should be involved in the process of identifying and defining the needs of the customers (stakeholders at large) and how to meet these needs in the most cost-effective manner.* Marketing-led organisations pursue this approach called Total Quality Marketing (**TQMar**) [Witcher, 1990].

A TQMar organisation's mission statement need to be changed from [Witcher, 1990]:
"Supplying value satisfactions at a profit"
to:
"Supplying optimal value satisfaction to the customers at optimal profit and optimal job satisfaction whilst improving the physical and social environment".

A marketing oriented organisation considers its customers in all its workings and operations. Making this happen within the company is the prime role of modern marketing. This requires more than a simple marketing organisation which oversees and co-ordinates only those activities directly involved with the external customers -- such as sales, promotion and delivery. The implementation of the marketing concept cannot be solely confined to a marketing department. The co-operation of senior management in every function of the company is important. Every employee influences, and is influenced by organisational climate and culture. The greatest determinant is the company policy and the quality of the company's top management.

The marketing concept goes far beyond market segmentation and targeting. There are three related conditions that a company's management must try to fulfil if the company is to be truly marketing oriented. These are:
- **Condition 1:** The company's strategy must be externally focused on customers in all aspects of organisation and operation.
- **Condition 2:** The company's operations must be co-ordinated and matched with the needs of target customers.
- **Condition 3:** Everybody who works for the company must participate in a total marketing environment.

An example of a company going for TQMar is shown in Mini-Case 5.4.

Chapter 5: Business Process Re-engineering (BPR)

Mini-Case 5.4:

British Telecom Goes into TQMar   [Witcher, 1990]

In 1987, British Telecom (BT) introduced a full scale **TQMar Programme**. The convergence of telecommunications, computers, and consumer electronics technologies has led to new telephone product applications in final markets. These opportunities, coupled with the liberalisation of telecommunications and the subsequent privatisation of the company, have had profound implications for the corporate culture of BT. It had to move from an essentially engineering based company, with an administrative style of management, to one which is market-led, with a commercially responsive style of management. Changes in corporate culture are the most difficult thing for a large company to achieve. BT is a company of about 240,000 employees.

The company began the TQMar Programme in an attempt to commercialise its workforce from the top management down to the operational level staff. This programme involved several parts:

1. Spreading BT's mission, based on a full explanation throughout the company of end-user requirements from BT service. The mission statement incorporated customer service, product quality, the involvement of suppliers and attainment of a positive market position, the welfare of staff, and involvement in the community at large.
2. The use of cross-functional training and workshops, to involve participants from every level of the company.
3. A full discussion of the purpose of the Programme, its concept and need, how one's work relates to that of others, the techniques required to facilitate working together, and how to obtain commitment over time. These sessions involve a series of stages to guide discussion and work to suggest answers to two basic strategic questions: Where do we want to be?
      What do we want to achieve?
4. The determination of acceptable levels of failure, and review progress.

The emphasis of the TQMar Programme is on quality of service achieved through everyone rather than on the old supervisory techniques or quality control. The idea is to implement planning and management through the use of team work, and by so doing create an understanding which leads to a greater spread or responsibility, and proactivity. BT hopes these things will eventually improve its customer service and contacts and lead to a better corporate image.

To some extent these ideas are no more than an exercise of improving internal public relations. TQMar aims to making this process continuous by involving people who not only understand the reasons behind the scene but are eager to change their actions and accept new, proactive working habits.

A Total Marketing Corporate Environment

The marketing process consists of the decision on the marketing mix -- Product, Place, Promotion, and Price (4-P). Marketing is more concerned with the effectiveness (doing the right things) than with the efficiency (doing the things right) of a business. Therefore the success of marketing function should be the primary objective of the firm. Unfortunately, historically, TQM does not seem to start from marketing, but rather from production/operations. Therefore, there is an urgent need to look closely at what marketing can offer to TQM and vice versa

Tom Peters [1988] is one of the first writers who worked on bringing marketing and TQM ideas together. If quality is defined as fitness for use, then surely this must be done as a recognised part of internal marketing. Marketing texts and training practice must begin to take notice of the correspondence between quality, marketing and training. These are issues about how management and employees can be 'commercialised' in large companies, so that they take a holistic approach to marketing, which is true to the modern meaning of the name.

If 'total' ideas or schemes are to work for internal marketing, then the backing of top management is necessary. Marketing managers should be encouraged to inform others about market trends, competition, and their own company's market performance, to balance all of the company's operational functions in a way which pulls everybody in the same desired direction. For example, promotion and sales must not promise what production lines and delivery systems cannot produce, and it is the marketing function's job to provide the balance. It is insufficient to simply give people in non-marketing departments so-called marketing information. It is also necessary to increase their motivation and understanding -- to provide a context for using marketing information.

The TQMar Process

TQMar involves continuous improvement in the marketing functions. For successful implementation of TQMar and in order to achieve the required performance standards, performance measurement and monitoring and maintenance of appropriate levels of quality are necessary. To facilitate this necessity the following seven-step process is recommended:

Step 1: Clarify the vision in relation to market opportunities, where companies have had difficulties in understanding, developing, communicating and sharing common values, relative to their future growth and the involvement of their people.

Example: "What sort of business do we want to be?"

Step 2: Set tangible goals to express what the company wants to achieve, how the activities of all the people within the company contribute towards the total and the importance of the fact that the performance of all the individuals involved should be open to evaluation.

Chapter 5: Business Process Re-engineering (BPR)

Example: The low cost producer challenge
Cross product line opportunities

Step 3: Develop awareness and understanding, i.e., where the company is now, how the situation has evolved and what trends and options are open for the future.

Example: Comparative perceptions between customers and salesmen

Priority	Customers	Salesmen
1st	Quality	Price
2nd	Reliable Delivery	Quality
3rd	Rapid Delivery	Technical Advice
4th	Price	Product Range

Step 4: Maintain business activities to make sure that the strategies and tactics are mutually reinforcing and have been effectively communicated.

Example: Apply Deming's PDCA cycle in strategic planning and implementation.

Step 5: Set dynamic standards to create clear and consistent operating milestones which also remain relevant to the changing dynamics of the marketplace.

Example: Most companies work against budgets, few set dynamic standards. In order to set dynamic standards, use the 5-W and 1-H:

WHY	Should they be set?
WHERE	Should they be set?
WHAT	Should be measured?
WHO	Should be doing the measurement?
WHEN	Should measurement take place?
HOW	Should they be measured?

Step 6: Planned implementation designed to ensure that strategies and policies are put into practice through structured marketing planning systems and sequences, following through all aspects of the marketing mix and ensuring that contacts with customers are consistent at all levels.

Example: There need to be a deep understanding of the external environment with marketing activities being prioritised.

Step 7: Insist on creative evaluation, which does more than inspection and requires that the diagnostic monitoring capabilities and the inherent attitudes exist to enhance market-driven performance.

Example: The evaluation need to involve those who have an input to the results. The successfulness should be judged by market and external factors. This means that an increase of profit of say 20% over last year may be a bad sign, if the industrial average increase is 50%.

The TQMar Areas

The areas that TQMar is applicable includes the marketing data, resource, product, price, place, promotion, and sales activities. They are explained below:
1. TQMar - Data should address:

- Financial/historical/geographical details
- Sector applications and trends
- Purchasing criteria
- Distribution and selling methods
- Technological and environmental issues
- Competitive picture
- Relative perceptions
- Single European Market

2. **TQMar - Resource** should address:
   - Effective utilisation of current and fixed assets
   - As compared with other businesses in the industry?
   - Accuracy of product costing systems
   - Total management involvement in marketing planning
   - Level of integration of marketing and selling activities
   - The leading discipline in the business?

3. **TQMar - Product/Service** should address:
   - What is the customer buying?
   - What performance expectations do customers have?
   - What are the perceived benefits over the competition?
   - What level of quality is required?
   - What range and mix is needed?
   - What is the impact of the Single European Market?

4. **TQMar - Price** should address:
   - Consistent and logical policies
   - Process for change
   - Cost/profit relationships
   - Strategy cohesion
   - Value perception
   - Impact of the Single European Market

5. **TQMar - Place** should address:
   - How do your customers access you and trends?
   - What levels of service are customers demanding?
   - Which channel provides the right potential and quality?
   - How best to support the chosen channel?
   - Is there a system for performance control?
   - Which physical distribution standards are required?
   - What is the impact of the Single European Market?

6. **TQMar - Promotion** should address:
   - What is the product/service image?
   - What is the expected packaging standard?
   - Is advertising cost-effective?
   - Is the timing of promotion right?
   - Which are the target audience during a promotion campaign?
   - What is the most important message in promotion?

7. **TQMar - Sales** should address:

Chapter 5: Business Process Re-engineering (BPR)

- Level of integration of sales with marketing activities
- The sales person's familiarity with his customer's needs and wants
- The sales person's capability to meet the needs of his customer
- The amount of care the sales person provides to his customer
- Cost-effectiveness of the sales calls

Each of the above seven TQMar areas should be subjected to continuous improvements. An example of the importance of TQMar when applied to *promotion* is illustrated in the Mini-Case 5.5, which is followed by another example of organisation-wide application of TQMar (Mini-Case 5.6).

■■■

Mini-Case 5.5:

TQMar - Promotion: TV Advertising

All over the world companies marketing popular fast moving consumer goods spend a huge amount of money on TV advertising which is believed to be an effective medium for mass communication. Is it really effective? It depends entirely on the quality of the commercial. The difference between an eye-catching TV commercial and a mediocre one could spell success or failure.

Success of a TV commercial is usually measured by the 'recall rate' which is the percentage of the sample population who can recall a piece of advertisement after a certain time period. There are some professional institutions that survey recall rates of TV commercials.

In the UK for instance, during 1992, Nescafe Gold Blend (promoted by McCann-Erickson) has been on the top of the league table researched by the Chartered Institute of Marketing for six consecutive months [CIM, 1992]. The reason for such a remarkable success was very simple. McCann-Erickson used the most common human nature -- being nosy. People generally like stories, especially when given the opportunity to guess or invent the ending themselves. So that was the trick.

Nescafe Gold made up the story evolving between a young, prosperous male and a charming, professional female, who used the Nescafe as an excuse to meet one another. The advert has been made of several short but catchy scenes.

Every time when Nescafe Gold was on air, people were curious about what was going to happen with the couple next. This is human nature, though. The Nescafe advert became the topic of discussion among friends. It made such a big impact on people that no one forgot to pick up Nescafe Gold while doing his weekly shopping in a supermarket, .

Ever since this success, there has been a trend in the TV commercial world to run stories. There are various advertisements using similar tactics by companies such as Renault, Rover Cars, British Meat, and so on. Recently, even the advertising giant MacDonald's ran a story with its restaurant as a meeting place for a divorced couple tactfully organised by their son.

As the famous comedian once said in his programme "Canned Carrot -'The world's best TV commercials", people like watching TV commercials rather than the programmes because they are simply better. Effectiveness in a TV commercial, despite its cost, lies in its quality!

■■

■■

Mini-Case 5.6:

Lucas Industries achieving success through TQMar [Wilson, 1989]

Lucas Industries plc is one of the largest industrial groups in the UK. During 1985-1988, it suffered sales down by 35%, profit down by 16% and return on asset (ROA) up by 30% (due to sell off of assets). In 1988, it launched the TQMar approach to re-engineer the business. It first did a marketing audit and findings of the Do-s and Don't-s were:

Market Analysis	Do-s	Don't-s
Quantitative:	☑ Existing markets	☒ Key global markets ☒ Market profitability
Qualitative:		☒ Why to buy? ☒ How to buy? ☒ How to choose?
Segmentation:	☑ Existing sales by product	☒ By market ☒ Global
Goals and Strategies:	☑ What?	☒ How? ☒ Differentiation ☒ Positioning
Distribution:		☒ Channel strategies ☒ Distributor development
Key Account Management:	☑ Some	☒ Profit analysis ☒ Documentation ☒ Pan-Euro accounts
Marketing Mix:	☑ Sales activity	☒ Planning ☒ Control ☒ Commonalty ☒ Strategy fit

Chapter 5: Business Process Re-engineering (BPR)

As a result of the above marketing audit, Lucas agreed on the definitions of TQMar in the areas of Strategies, Processes and Systems. Then they arrived at the following deliverables:
- *Universal process -- company-wide manual issued.*
- *Trained people -- marketing management workshops*
- *Marketing strategies -- Strategic Business Units (SBUs) established:*
    - Automotive -- Europe
    - Automotive -- Overseas
    - Industrial -- Global
    - Electronics -- Global
    - Government -- Global
- *Marketing plans -- business unit marketing plan*

One year later:
- *Mission statement announced*
- *Processes in place*
- *Management training done*
- *Operating manuals in use*
- *Marketing plans prepared*
- *Customer care initiative started*
    - Performance: Sales up 18%
    - Profit up 10%
    - ROA up 5%   (a real increase)

■■■

Total Quality Purchasing

Purchasing's responsibility with regard to the quality of purchased materials should not be defined too narrowly. Product development can generally be classified into pre-development and post-development stages. Purchasing should be involved in the pre-development stage of new products or projects in order to investigate at an early stage the extent to which purchased parts required for end products are available in the market. Early involvement of purchasing will result in suggestions for product, materials, and/or design alternatives. Purchasing should also ensure that goods ordered are delivered according to specifications in the post-development stage of new products or projects. Thus **Total Quality Purchasing (TQPur)** can be defined as *the organisation-wide participation in the improvement of the logistic process to meet the needs of the customers.*

In general, in manufacturing firms, over half of the cost of the final product comes from purchasing. In terms of the supply chain concept, purchasing has been considered as an integral part of the quality function. On the other hand, purchasing is mostly considered as a low profile activity for services organisations. In fact, purchasing in services organisations is of *no less* importance than their manufacturing counterparts, if one considers the recruitment of human resources as part of the purchasing function.

The scope of the purchasing function can be characterised by three alternative models:
- Purchasing is considered by management to be a clerical function.
- Purchasing is considered to be a significant commercial activity, handling a large part of the company's material and service expenditures.
- *Purchasing is considered by management to be a strategic business function.*

Depending on the view, the position of the purchasing department within the organisation and the measures used for purchasing evaluation will differ significantly as shown in Figure 5.7.

Scope of purchasing	Position of purchasing	Purchasing Performance Measures	Focus on
Clerical function	Low	Number of orders, backlog, purchasing administration, lead time, authorisation, procedures, etc.	efficiency
Commercial function	Report to management	Saving, cost-reduction, negotiation, contracting, single/sole sources, etc.	efficiency
Strategic business function	Integrated in strategic planning	Supplier development, make vs. buy studies, integration with R&D, value analysis, purchasing engineering, etc.	effectiveness

Figure 5.7 How management looks at purchasing

Why Measure Purchasing Performance?

The benefits derived from a systematic purchasing performance evaluation are:
- Purchasing performance evaluation can lead to better decision making, since it identifies variances from planned results; these variances can be analysed to determine their causes and action can be taken to prevent them in the future.
- It may lead to better communication with other departments: e.g., analysing payment conditions with financial management and deciding on specific payment procedures may improve mutual understanding; establishing an effective materials budget requires co-ordination with production and inventory control -- a team effort; etc.
- It makes things visible: regular reporting of actual versus planned results enables a buyer to verify whether his expectations have been realised. This

Chapter 5: Business Process Re-engineering (BPR)

provides constructive feedback to the buyer and it also provides information to management about individual and group effectiveness.
- It may contribute to better motivation: properly designed, an evaluation system can meet the personal and motivational needs of the buyer. It can be used effectively in a constructive goal setting, motivational, personal development programme.

In summary, these benefits indicate that purchasing performance evaluation should result in a higher value added by the purchasing department to the firm. This higher added value might take the form of operating cost reductions, lower material prices, fewer rejects, better sourcing decisions, etc.

What Should Be Measured?

Four key areas in purchasing performance evaluation can be identified:

Purchasing Effectiveness
- Purchasing Price / Cost Dimension
  -- materials costs / prices
  -- competition
  -- co-ordination
  -- transportation .....
- Purchasing Quality Dimension
  -- specifications
  -- rejects / returns
  -- purchasing engineering
  -- rework
  -- production stops .....
- Purchasing Logistics Dimension
  -- lead times
  -- order-quantity
  -- inventories
  -- vendor-performance
  -- deliveries (late, on time, early)...

Purchasing Efficiency
- Purchasing Organisation Dimension
  -- workload / orders / quotations
  -- procedures
  -- information system
  -- management
  -- expertise .....

The discussion of TQPur would be incomplete without mentioning the implications on purchasing evaluation of the just-in-time (JIT) production concept and the explosive growth of computer-based systems. Utilisation of the JIT philosophy requires, among other things, a supporting supplier network that is able to meet stringent quality and delivery requirements. Both factors are crucial, since firms operating under a JIT production system do not have significant inventories that can buffer against unreliable supplier performance. Further, since inspection of incoming goods is minimised, each defective item purchased can affect production continuity. These considerations amplify the need for current evaluation of supplier performance.

Since an effective purchasing function is vital to the success of the JIT concept, most observers believe that management will support even more strongly the development of effective performance measurement and evaluation programmes in the purchasing area. Finally, all of these considerations are enhanced substantially by the astounding developments occurring in the computer industry. Computer hardware and software is now becoming readily available for use in operating the type of measurement and evaluation system discussed here. The on-line, interactive capability of current information systems greatly facilitates day-to-day operations of the evaluation programme.

At this point, it is worthwhile mentioning that JIT production system can be applied to the production of services. Many banks, delivery services, and other service industries are achieving success with the JIT concepts. More references can be found in [Ho, 1988]. Furthermore, an example of application of TQPur can be found in the Mini-Case 5.7.

■■

Mini-Case 5.7:

Total Quality in Purchasing   [Wong, 1995]

In 1994 Wong surveyed purchasing professionals in Hong Kong business organisations in order to find out the TQPur practices. There were 314 questionnaires sent to the purchasing managers of publicly listed companies and 673 questionnaires were sent to the members of the Hong Kong Institute of Purchasing and Supply. The questionnaire consisted of four parts: the first part asked the respondents for basic information concerning their companies and the purchasing department; the second part tapped respondents' view on their orientation to quality; the third part aimed to assess respondents' opinions towards the results of TQPur; the last part to find out from the respondents the present practices between their purchasing department and their suppliers. The 193 responded questionnaires resulted in a 20% response rate. The results are summarised as follows:
1. Information about the company and purchasing department
Most of the companies were in the medium to large size category (above 500 employees). 38% of the companies have adopted a TQM system. Regarding the reasons for the adoption, the most frequently chosen response was higher quality demands by customers (78%), followed by quicker response to the market (47%) and quality is a selling point (40%).
2. Orientation to Quality
In determining the award of business to suppliers, respondents were asked to rate the importance of five factors on a scale of 1 (most important) to 5 (least

Chapter 5: Business Process Re-engineering (BPR)   129

important). The respondents ranked quality as the most important factor with an average of 1.45, followed by price .78, reliability 1.85, delivery 2.03 and service 2.17. Moreover, the analyses revealed that 66% of the respondents were willing to pay higher prices for higher quality materials.

3. Results of TQPur

Respondents were asked to express on a 5-point scale their views of the results of quality of materials purchased. The responses indicated that they considered quality materials as important in achieving 'good quality product', 'quicker delivery time', 'less inventory' and 'higher profit'. Moreover most respondents circled Strongly Agree for 'good quality product'.

4. Present practices with suppliers

Respondents were asked to give information about present practices in the relationships with suppliers. More than half of the respondents are practising the basic features of a quality approach to suppliers which include: multiple sourcing, obtaining suggestions from suppliers for quality improvement, sharing information with suppliers, involving suppliers in product design, reducing the number of suppliers, and partnering with some suppliers. On the other hand, a relatively smaller percentage practise the more advanced features of partnership relationship such as: giving quality awards to suppliers (26.4%), single sourcing (30.5%), giving quality training to suppliers (33.7%), and requiring supplier certification (34.7).

5. Differences between TQM companies and non-TQM companies

The TQM companies have more faith in the benefits of quality purchasing. Furthermore, a higher proportion of TQM companies (50%) have applied the more advanced features of the quality approach compared with the non-TQM companies (20%).

The responses to this exploratory survey indicate that, despite the severe competitive nature of doing business in Hong Kong, the purchasing professionals put more emphasis on quality than price. This takes the trend away from the past practice of awarding business on price, and leads towards a new quality orientation.

■■

Influenced by the TQM approach, TQPur has to adopt a new strategy of operation with emphasis on quality rather than price. In addition, there is a trend towards supplier partnering relationships. In the selection and evaluation of suppliers, quality has been perceived the most important factor, while other considerations such as price, have become less important [Hutchins, 1992]. Moreover, purchasing professionals are pursuing practices to improve supplier relationships and ensure quality. These entail supplier certification, reducing the supplier base by moving towards single sources of supply, investment in the training and support of suppliers, involving suppliers in the product design, and the sharing of information and rewards with suppliers.

Why is BPR useful?

Improvements in business performance of, say, 10-15 per cent can be achieved in most companies using conventional consultancy techniques. Where quantum leaps are required -- for example, where the old needs to be completely replaced with the new -- then re-engineering is a good way forward. The key to grasping the way BPR differs from other improvement studies lies in understanding the focus, breadth and duration of the re-engineering process.

The primary focus is on the customers -- those people who pay the money which keeps the business going. So if a process does not help to serve a customer then why have the process in the first instance? Although BPR requires a detailed knowledge of what the customers want it does not demand a highly detailed understanding of the tasks involved in every activity of the business. This makes BPR economical in terms of investigation time when compared with conventional methods, in which highly-detailed studies are usually undertaken before any change is made. BPR requires that those conducting the study are highly experienced in business practices and systems, and are able to identify the features of the business which are crucial to its success. A high-level in-house team, working with experienced consultants, would be able to provide the necessary expertise.

A further facet of the BPR approach concerns the speed with which changes are introduced. Conventional wisdom states that change is best brought about through an evolutionary approach. If it is required to introduce a radically changed organisation, it can be argued that it makes good sense to carry out the necessary changes quickly. Many major BPR projects have been implemented within one year [Ovenden, 1994]. Another example of successful application of the BPR concept is given in the Mini-case 5.8 below.

■■

Mini-Case 5.8:

Boots The Chemists Goes for BPR   [Peratec, 1994]

Boots the Chemists is the largest chemists retail chain in the UK and has branches in Europe and other part of the world. Unique among UK multiple retailers, Boots has maintained a substantial inhouse facility to provide and maintain the company's infrastructure. The Boots Store Planning Department has a staff of 350, responsible for all of the company's fixed assets. Its activities include creative design, planning, project management and implementation of a substantial capital development programme, store cleaning and a facilities management service for a property portfolio in excess of 2000 sites.

Before BPR implementation, the Department faced a major problem -- a reputation for being 'big, slow, expensive and inscrutable'. The Department's

Chapter 5: Business Process Re-engineering (BPR)

senior management team saw BPR as a vehicle for change. After analysing their business in great depth, they built up a step change improvement.

The Department first identified and traced the key processes across the business of Boots. This analysis highlighted the need for simplification in store planning processes. Prime goals were improved customer focus, strengthened project management, the creation of performance measures and regular benchmarking -- all to improve the bottom line.

Further contributions to the case for change were made by an analysis of the total Cost of Quality -- the cost of quality conformance plus the cost of non-conformance, such as the abortive work, errors and rework. This amounted to a staggering £2.5m, or 25% of Department's cost base, and 17% of this cost was due to non-conformance.

Having defined these goals and identified the key value-adding processes needed to reach its targets, the Department restructured its entire operation. A Store Design Concept for clients has been defined and standard 'models' applied in place of expensive, bespoke technical solutions. **Project management processes** *have been established, a post-scheme* **'after-care' system** *set in place and* **cost control** *re-evaluated.*

Some of the major achievements attributed to the BPR initiative are:
- *The department has relocated, and reduced its size by 13%.*
- *There are now fewer management layers within the department.*
- *Staff are empowered through Continuous Improvement Teams.*
- *The Department's services are now 'indispensable' and 'value-adding' according to its customers.*
- *The infrastructure is now in place to ensure continuous improvement.*

■■

How to Implement BPR

Organisations will avoid the problems of 'change programmes' by concentrating on 'process alignment' -- recognising that people's roles and responsibilities must be related to the processes in which they work. Senior managers may begin the task of process alignment by a series of BPR steps that are distinct but clearly overlapped. *This recommended path develops a self-reinforcing cycle of commitment, communication, and culture change.* The steps are as follows.

Step 1: Gain commitment to change through the organisation of the top team.
Step 2: Develop a shared vision and mission of the business and of what change is required.
Step 3: Define the measurable objectives, which must be agreed by the team, as being the quantifiable indicators of success in terms of the mission.
Step 4: Develop the mission into its Critical Success Factors (CSFs) to coerce and move it forward.

Step 5: Break down the CSFs into the key or critical business processes and gain process ownership.
Step 6: Break down the critical processes into sub-processes, activities and task and form the teams around these.
Step 7: Re-design, monitor and adjust the process-alignment in response to difficulties in the change process.

<pre>
             Step 1: Gain top level commitment
                              |
                              V
          Step 2: Identify mission and change required
                              |
                              V
             Step 3: Define measurable objectives
                              |
                              V
              Step 4: Develop mission into CSFs
                              |
                              V
           Step 5: Develop CSFs into critical processes
                              |
                              V
        Step 6: Develop critical processes into sub-processes
                              |
                              V
            Step 7: Re-align process if problems arise
</pre>

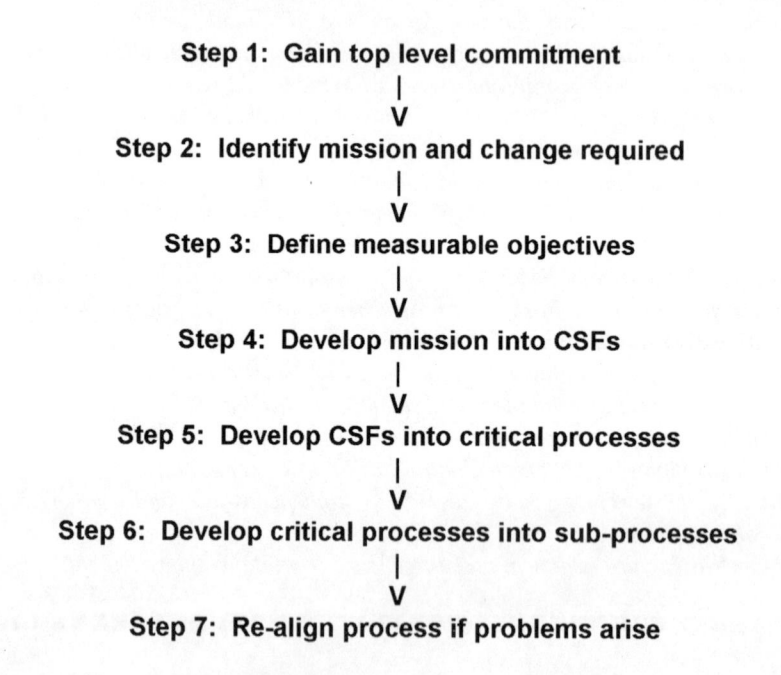

Figure 5.8 Top-down Approach to Business Process Re-engineering

The seven steps are summarised in a diagram as shown in Figure 5.8. BPR creates change. Change must create something that did not exist before, namely a **'learning organisation'** capable of adapting to a changing competitive environment. A learning organisation aims to create a self-perpetuating momentum which changes the culture of the organisation. That is to say, the aim is that the norms, values and attitudes underpinning behaviours be changed towards continual questioning and continual improvement. It embraces human resources development on the one hand, and systems development (including BPR) on the other. For without addressing the systems of an organisation, from communication and information systems to reward and recognition systems, you are building your houses on sand, foundationless.

The organisation must also learn how to continually monitor and modify its behaviour to maintain the change-sensitive environment. Some people will, of

Chapter 5: Business Process Re-engineering (BPR)

course, find it difficult to accept the changes and perhaps will be incapable of making the change, in spite of all the direction, support and peer pressure brought about by the process alignment. There will come a time to replace those managers and people who cannot function in the new organisation, after they have been given the opportunity to make the change.

With the growth of people's understanding of what kind of managers and employees the new organisation needs, and from experience of seeing individuals succeed and fail, top management will begin to accept the need to replace or move people to other parts of the organisation. An example of such organisational change is given in Mini-Case 5.9.

■■

Mini-Case 5.9:

BPR at Co-op Bank -- Improving Personal Customer Service   [Dignan, 1995]

The Co-operative Bank strategy of encouraging its customers to telephone or write to its account management centre, rather than the branches, proved successful. The Co-operative Bank has over 100 branches across the UK, and the country's largest telephone banking operation, serving the needs of its 1.5 million account holders. It is a member of the LINK organisation which provides a network of over 6,000 cash dispensers nation-wide. The bank has long had a reputation for innovation with many of its products and services being "first" on the UK banking scene.

Volumes grew rapidly, placing pressure on service quality resulting in lost telephone calls, correspondence backlogs and duplication of effort as work was handed off to other areas. A project team was formed to re-engineer fundamentally business processes and to realign the organisation, leading to improvement in customer service. Account Management Centre employs some 600 staff, in an operation accessible by telephone 24 hours a day 365 days per year. It receives over 4 million calls, and processes over 5 million pieces of paper a year. The project commenced in 1993, looking at training, process improvement and control, customer care, account opening, work-flow management, benefits tracking, communication and infrastructure. The project was concluded in May 1994 when the last of 34 multi-skilled customer-service teams pulled out. Recent research found that among high street banks Co-operative Bank had the highest customer satisfaction rating scoring 94% among its customers.

■■

Chapter Summary:

- BPR is the radical redesign of business or individual work processes in order to maximise its effectiveness. Process owners are charged with ensuring that employees understand how their individual work processes affect customer satisfaction.
- Service industry and quality -- The competitive advantage in the service organisation lies in its quality leadership in the market place. The challenge for the organisation is to perform at a level that meets all expectations. SERVQUAL application in educational institutions has been used as an illustrative example.
- Best Practice Benchmarking (BPB) -- BPB is a technique used by successful companies around the world to help them become as good or better than the best in the world in the most important aspects of their operations.
- TQMar -- Is about how to supply optimal value satisfaction at optimal profit and optimal job satisfaction whilst improving the physical and social environment.
- TQPur -- Purchasing is considered by management to be a strategic business function.
- Usefulness of BPR -- Where quantum leaps are required -- for example, where the old needs to be completely replaced with the new -- then BPR is a good way forward.
- Implementation of BPR -- The recommended path is to develop a self-reinforcing cycle of commitment, communication, and culture change through the development of a learning organisation.

Hands-on Exercise:

1. Apply the SERVQUAL gaps to your job and then your organisation. Are there any gaps which require your immediate attention?

2. Develop a TQMar process for your own organisation. What are the potential benefits?

3. Develop a TQPur process for your own organisation. What are the potential benefits?

4. Following the seven step procedure in Figure 5.8 formulate a plan to implement BPR in your own organisation.

Chapter 6

Quality Control Circles & Problem Solving

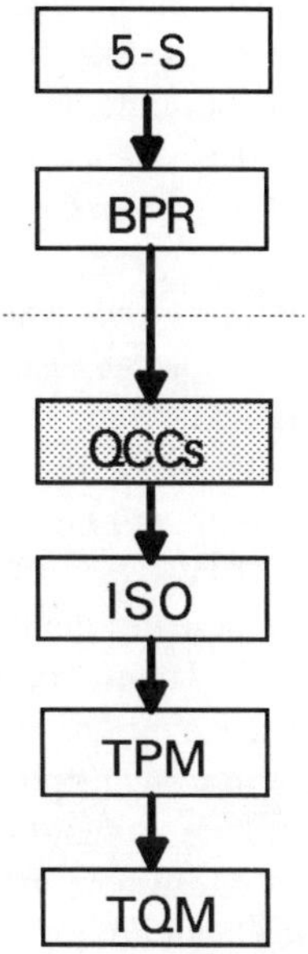

Topics:

- What is a QCC?
- The Japanese QCC experience
- The seven quality control tools
- Problem solving
- The S-S problem solving method and examples
- The S-S Method and the seven QC tools
- Why do we need QCCs?
- How to establish QCCs
- QCCs and employee empowerment

```
5-S
 ↓
BPR
 ↓
QCCs
 ↓
ISO
 ↓
TPM
 ↓
TQM
```

Introduction: What is a QCC?

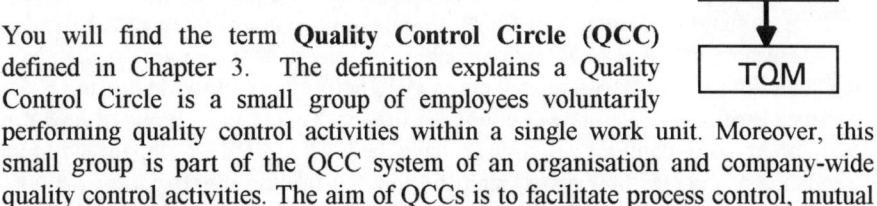

You will find the term **Quality Control Circle (QCC)** defined in Chapter 3. The definition explains a Quality Control Circle is a small group of employees voluntarily performing quality control activities within a single work unit. Moreover, this small group is part of the QCC system of an organisation and company-wide quality control activities. The aim of QCCs is to facilitate process control, mutual and self-development of employees, and improvement of their workplace by utilising quality control techniques and full participation of all members.

Voluntary teams are not always called QCCs. Other names commonly used are: Quality Circle, Support Team, Synergy Team, Work Improvement Team, Action for Customer Team, or any other acronym which the team can identify itself easily. All QCCs follow an essentially standard pattern of approach to problems, which enables experience and enthusiasm to be passed on within and across company boundaries and between countries. Conferences of hundreds of active QCC members are now a regular occurrence in the UK (organised by the National Society for Quality through Teamwork, NSQT) and in every other country where QCCs have been established. The basic concepts behind QCC activities as part of the company-wide quality control effort are:
- To contribute to the improvement and development of the enterprise;
- To respect humanity and to build worthwhile lives and cheerful work areas;

- To give fullest recognition to human capabilities and to draw out each individual's infinite potential.

The Japanese QCC Experience

QCCs originated from Japan in the 1950s. No discussion of QCCs in Japan will be complete without mentioning the role played by the Union of Japanese Scientists and Engineers (JUSE), a private organisation set up in 1946 with the aim of studying foreign technology that might be applied to raise the level of Japanese industrial technology. In 1949, JUSE began organising Quality Control (QC) seminars which, through their frequency and expanding reputation, started attracting representatives from a large number of enterprises. Since then many seminars and lectures by prominent QC experts have been arranged. Mr. Ichiro Ishikawa, father of Prof. Kaoru Ishikawa -- the quality guru whose contribution to the quality movement is explained in Chapter 2, was the first chairman of the Board of JUSE. He took the initiative to organise seminars and invited Dr. Deming and Dr. Juran to Japan to give talks on QC and QC management during the 1950s (see Chapter 2 for details). These seminars were well attended by Japanese top managers.

In Japan every November is designated as the 'Quality Month' across the nation, and QC and QCCs are promoted and emphasised. In no other country is the movement promoted so continuously, nor is there any other country where 4 levels of annual QC conferences for presidents of companies, managers, foremen, and consumers are held so regularly. These are the general features of Japanese QCC, but do not necessarily represent the QCC activities of individual firms.

Since Prof. Kaoru Ishikawa established the first QCC Conventions in Japan in 1963, about 200 conferences a year have been organised in Japan, with an attendance of over 50,000 people and the number of case reports of over 4,000. There are over 1 million QCCs registered with JUSE, each producing an average of about 50 suggestions per annum. The number of non-registered QCCs is estimated to be 3 times as much. It is this driving force that leads Japan towards a nation of quality [NPB, 1984].

QCC Organisation Structure

In many Japanese organisations QCC activities are taken seriously. This is why the organisational structure incorporating QCC activities is elaborate and well defined. Each level of the organisation is normally involved in QCC implementation, usually starting from the production line. Established committees like promoter groups, leader groups, and promoter-leader groups are not uncommon. These groups or committees are important to the understanding of existing problems and their rectification. The committees also plan QCC activities for the future. In other words, these committees from the various levels help to monitor the progress of QCC activities.

Chapter 6: QCC & Problem Solving   137

Suggestion Schemes

Suggestion schemes are an important feature of the QC system in Japanese organisations. Suggestions may be submitted as individual or group efforts. QCCs that have completed their projects submit their suggestions to a Committee which reviews and grades them. Suggestions are usually rewarded with cash payments. Suggestion schemes play an important role in problem solving. Individual suggestions are usually considered minor as compared with group ideas. Certain organisations use notices to remind workers to continue putting in suggestions to the target of (say) 10 suggestions/head/year.

In-house QCC Presentation

In-house QCC presentations are often held to help sustain QCC activities. In most cases QCCs from sections make their presentations to section managers. The best QCC from each section is then selected to do a presentation in the department wide competition. The best presenting team here is then selected to present in the plant-wide competition. Sometimes, as a further step, the best QCC at the plant-side level is asked to make a presentation to the president of the organisation. Winning QCCs at each level are awarded plaques, certificates of participation or other mementoes (see Figure 6.1). It is not common for QCCs to receive monetary awards in the in-house convention. Monetary awards are usually given when a circle's suggestions have been passed to the Suggestion Committee.

Figure 6.1 QCC requires Recognition

The Seven Quality Control Tools

The Seven QC Tools were developed by Prof. Kaoru Ishikawa, 'the father of QCCs' [Ishikawa, 1982]. Through his experience, he identifies a set of tools which can be used by teams and individuals to interpret the data fully and derive the maximum information out from it. These are seven effective methods which will offer any organisation means of collecting, presenting, and analysing most of its data and problems. *They are: Process Flowchart, Check Sheet, Histogram, Pareto Analysis, Cause & Effect Analysis, Scatter Diagram and Control Charts.* Since their applications are so wide, they are also known as the **seven tools for QCCs.**

TOOL #1: Process Flowchart

A *Process Flowchart* (Figure 6.2) is a picture of what is actually done, using symbols as a mode of presentation. It records the series of events and activities, stages and decisions in a form which can be easily understood and communicated to all.

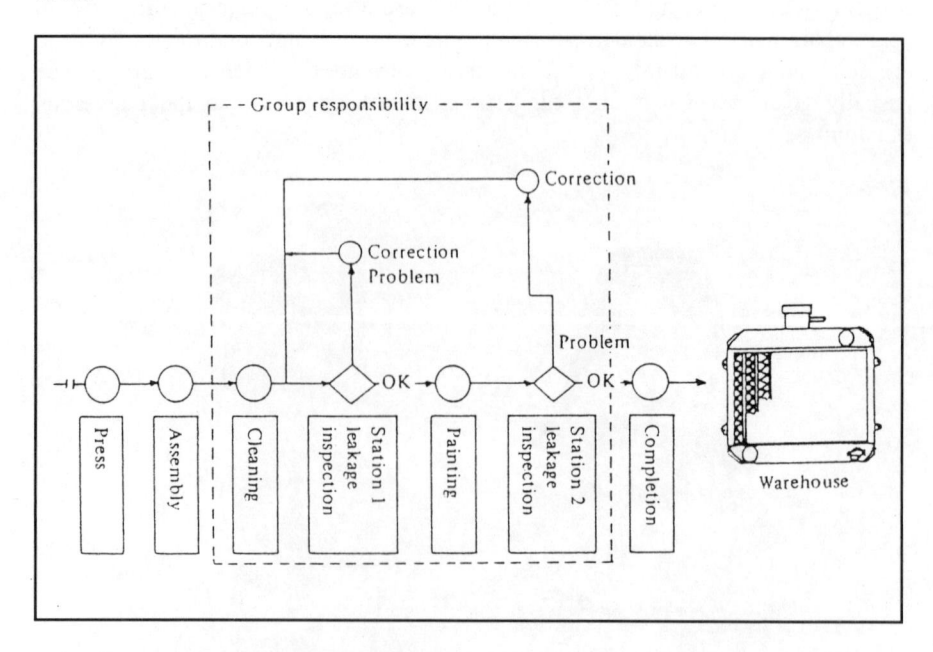

Figure 6.2 An Example of a Process Flowchart -- The work process of a work-group responsible for painting car radiators [Ishikawa, 1984]

By studying this chart and analysing its details, potential problems in the process can be spotted. This makes flowcharting a general problem-solving tool for

Chapter 6: QCC & Problem Solving 139

applications from administration to production processes. Supervisors and their teams use flowcharts to identify problems through the following steps. They get together to draw a flowchart reflecting the way the process actually works. After a consensus on this task, the team members are asked to write a flowchart on how the process should work ideally. The difference between the two charts represents the problems to be solved. It helps to understand the process and to make improvements.

TOOL #2: Check Sheet

A *Check Sheet* (Figure 6.3) or *Tally Chart* is a simple device on which data is collected or ordered, by adding tally marks against predetermined categories of items or measurements. It helps to structure data. For a situation to be correctly analysed and control to be realised, data must be collected carefully and accurately. Furthermore, the intent and purpose for which the data are collected should always be clear. Examples are: to control the production process, to determine the strength of materials, and to see the relationship between cause and effect of customer complaints. Of course, action should be taken when definite causes and effects become known.

CHECK SHEET		
Product: Hamburger		Date:
Stage: Final inspection		Shop:
Types of defect: undercooked, burnt, uneven colour, out of shape, cracked, others.		Inspector's name:
Total no. inspected: 1200		Lot no.:
Remarks: all items inspected		
Type	**Check**	**Sub-total**
Undercooked	‖‖‖ ‖‖‖	8
Burnt	‖‖‖ ‖‖‖ ‖‖‖ ‖‖‖ ‖‖‖ ‖‖‖ ‖‖‖ ‖‖‖	38
Uneven colour	‖‖‖ ‖‖‖ ‖‖‖ ‖‖‖ ‖‖‖	24
Out of shape	‖‖‖ ‖‖‖ ‖‖‖ ‖‖‖	17
Cracked	‖‖‖ ‖‖‖ ‖‖‖ ‖‖‖	19
Others	‖‖‖ ‖‖‖ ‖‖‖	14
	Grand Total	120
Total rejects	‖‖‖ ‖‖‖ ‖‖‖ ‖‖‖ ‖‖‖ ‖‖‖ ‖‖‖ ‖‖‖ ‖‖‖ ‖‖‖ ‖‖‖ ‖‖‖ ‖‖‖ ‖‖‖ ‖‖‖ ‖‖‖ ‖‖‖ ‖‖‖ ‖‖‖	88

Figure 6.3 An Example of a Check Sheet -- Defect item check sheet

In order to accomplish this efficiently, the data must be both easy to obtain and to use. This is why check sheets are important. Check sheets can be used for many purposes, but their most desirable characteristic is that they make it easy to compile data in such a form that they may be used readily and analysed automatically.

TOOL #3: Graphs

Most factories and businesses utilise many types of graphs at different points of production or work flow. There are numerous types of graphs ranging from a simple plotting of points to intricate graphic representation of complex and inter-related data. Basically, they are a tool for organising, summarising, and displaying data for subsequent analysis. To derive the maximum benefits, their purpose should always be known and their use periodically evaluated.

The most common types of graphs are the *Histogram, Line Graph, Radar-Chart and Pie-Chart*. Examples of application for each of these four types of graph are shown in Figure 6.4(a) to (d) respectively.

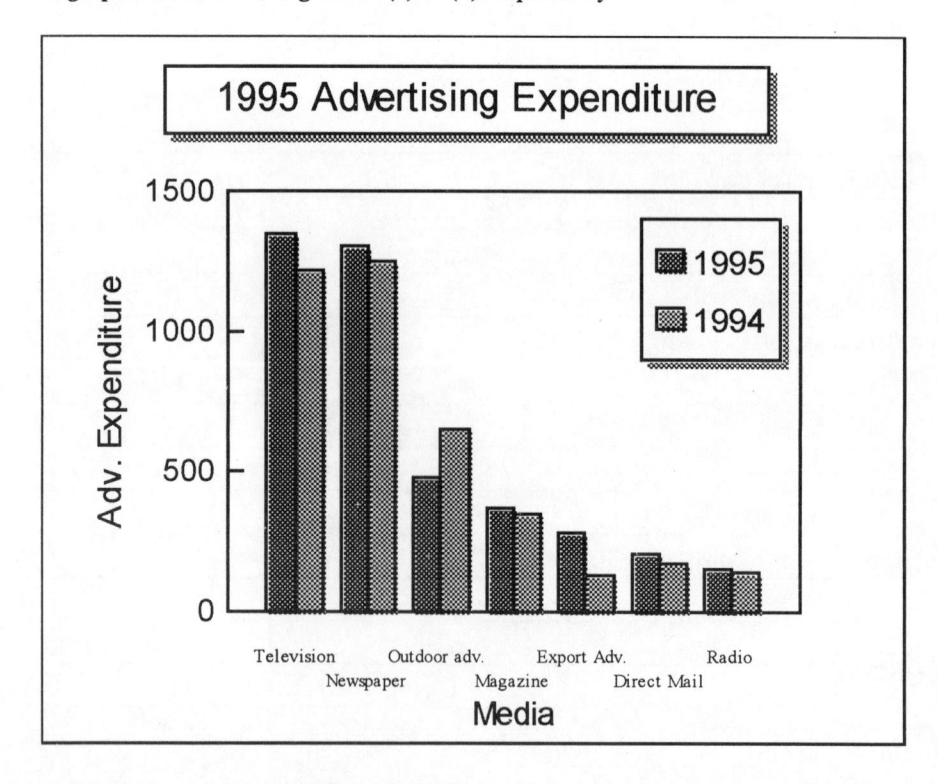

Figure 6.4(a) An Example of Graphs -- Histogram

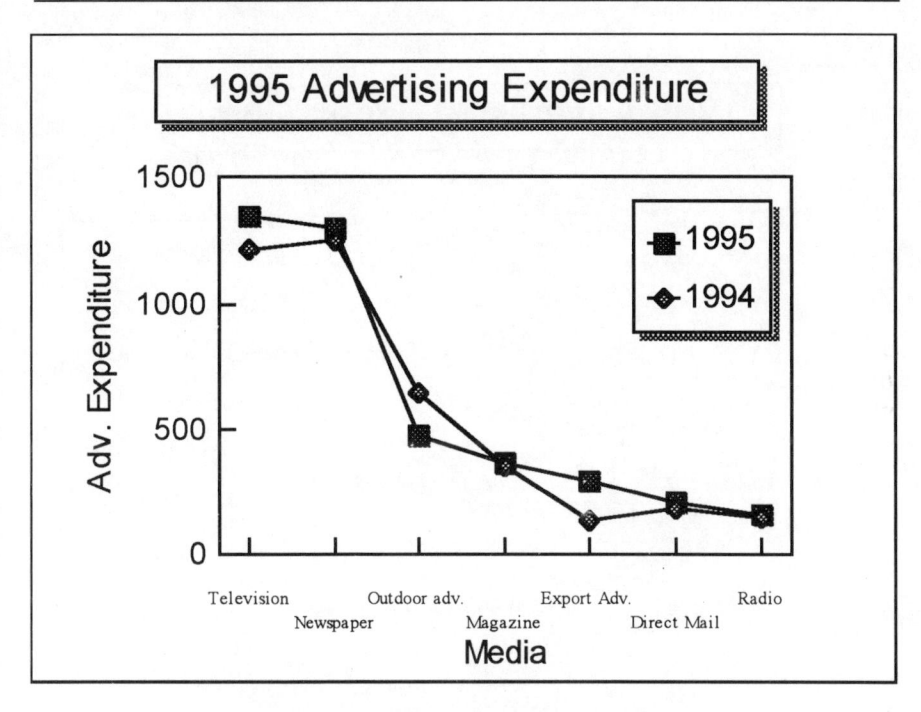

Figure 6.4(b) An Example of Graphs -- Line Graph

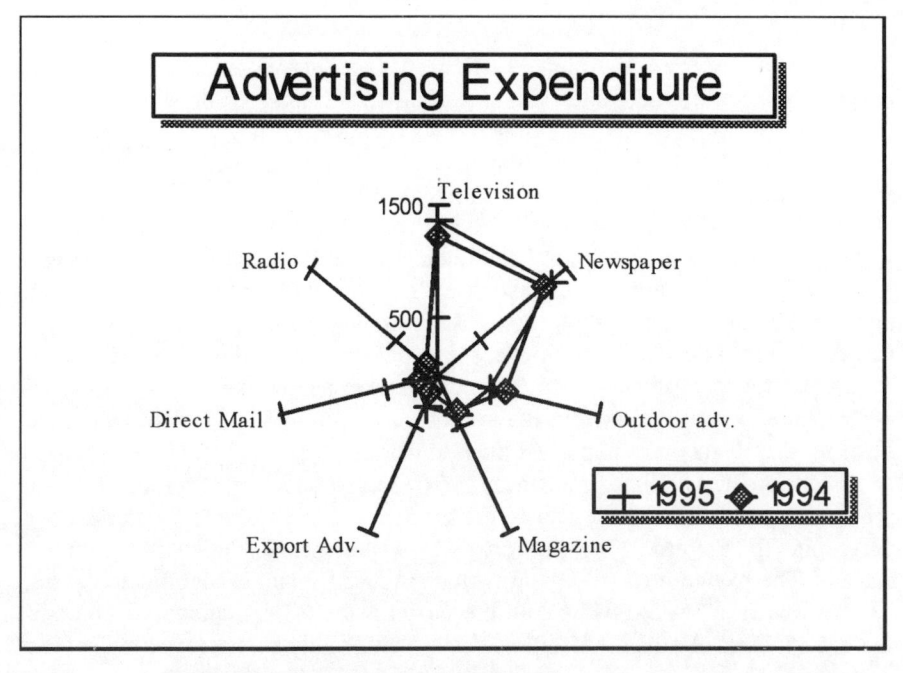

Figure 6.4(c) An Example of Graphs -- Radar Chart

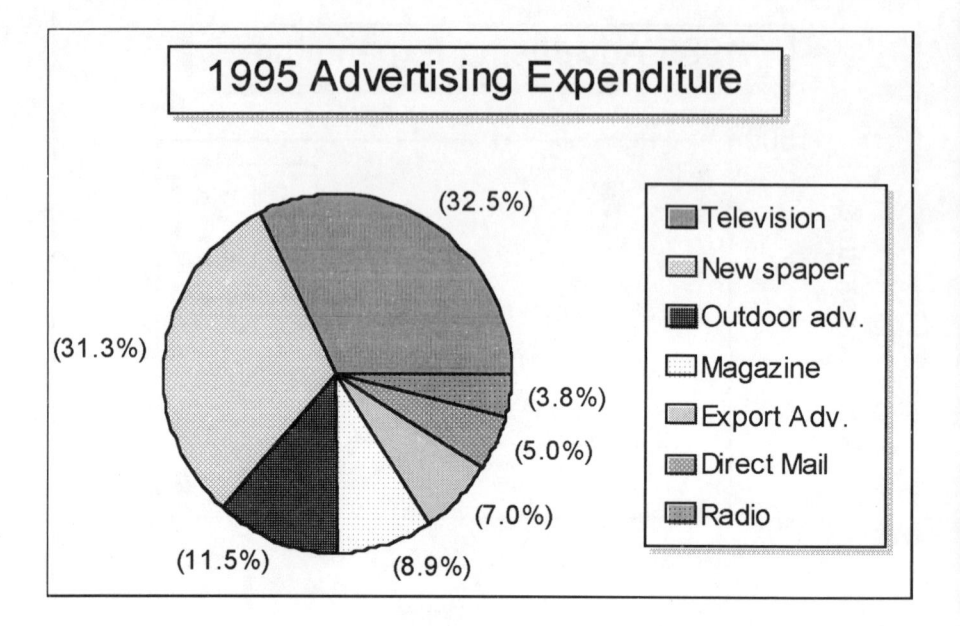

Figure 6.4(d) An Example of Graphs -- Pie Chart

TOOL #4: *Pareto Analysis*

In industrial processes and human activities, it is quite possible to identify the *causes* and quantify the *effects* resulting from each cause. If the significance of the effects are arranged in descending order, and the effects are plotted against the causes in the form of a histogram (Pareto Diagram), the first 'bar' will indicate the most significant cause, and so on. Pareto Analysis (Figure 6.5) is in effect a cumulative plot of this histogram, in the form of a line diagram.

Pareto Analysis is a picture of the frequencies of the 'vital few', or important items or causes of a problem, compared with the 'trivial many'. It is a bar graph and a line chart. The bar graph lists in descending order the problems affecting a process. The line chart accumulates the percentage of the total number of occurrence for each problem area. This gives rise to another name for this tool - **80-20 Rule**, indicating that 80% of the problems stem from 20% of the causes. It helps to identify the most important areas to work on.

Pareto Analysis is also known as **ABC Analysis.** If two vertical lines at around 20% and 50% of the causes are erected, they will divide the cumulative curve into three areas. The first area (A) would include the most significant causes. The second area (B) would include the next group of significant causes. The third area (C) would include all the 'trivial many'. Thus, attention should be focused on Area A, then Area B. Area C can usually be neglected or actions postponed.

Chapter 6: QCC & Problem Solving 143

Although Pareto Analysis is 'as easy as ABC' to use, its applications are sometimes beyond expectations. In fact, any management process requiring prioritisation with the effect that can be measured quantitatively will find application with Pareto Analysis. Examples are: identify the main cause of defects, delay in delivery, customer complaints, wrong data entry and calculation, equipment failures, billing errors, stock shortages, reservations not honoured, purchase order error due to incomplete description, warranty claims and late reports. Many a time, the benefits do not just come from the analysis phase, but also from the data collection phase. This is because the method encourages identifying all the possible causes and their numerical contribution to the effects. Once the data are collected, the process relationship becomes much clearer.

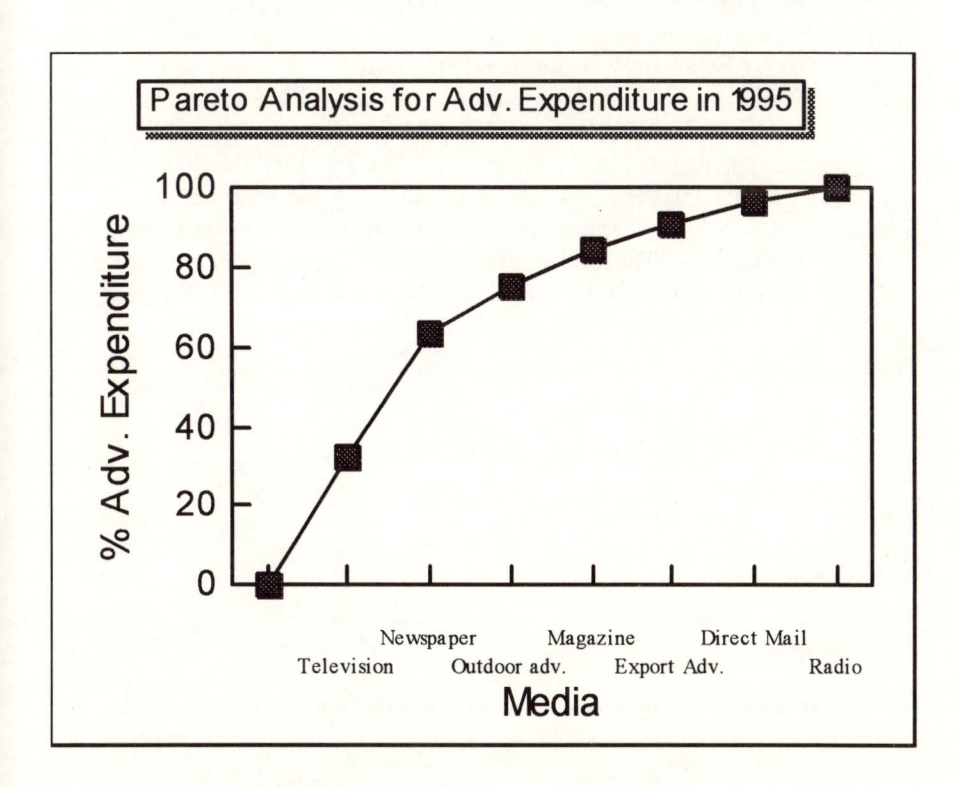

Figure 6.5 An Example of Pareto Analysis -- Cumulative curve of the Pareto Diagram on % Advertising Expenditure

'A' Class -- Television & Newspaper account for 63% of Total
'B' Class -- Outdoor & Magazine account for 21% of Total
'C' Class -- Others account for 16% of Total

TOOL #5: *Cause-and-Effect Diagram*

Cause and effect analysis and brainstorming involve the use of the expert knowledge of a group of people at work to identify certain **effects** by brainstorming and present their ideas as **causes** on a picture. The diagram resulted is called *Cause-and-Effect Diagram* (Figure 6.6). Because of its appearance it is also called *Fishbone Diagram*, the head of which is the 'effect' or problem, and the ribs of which carry the categorised possible causes. It is also called *Ishikawa Diagram*, owing to its inventor, Kaoru Ishikawa.

A Cause-and-Effect Diagram helps to open up thinking in problem solving. It is useful in sorting out the causes of dispersion and organising mutual relationships. There are three steps in drawing up the Cause-and-Effect Diagram:

Step 1: Determine the quality characteristics and write them on the right side, with a broad arrow going from left to right.

Step 2: Write the main factors which may be causing the **effect**, directing a branch arrow to the main arrow. It is recommended to group the major possible **causes** into such items as 5-Ms, i.e., Man, Method, Machine, Material and Money. Each individual group will form a branch.

Step 3: Onto each of these branch items write in detail factors and sub-factors, if necessary. The team leader must make sure all the items that may be causing the **effect** are included. Team members must discuss openly any possible ideas.

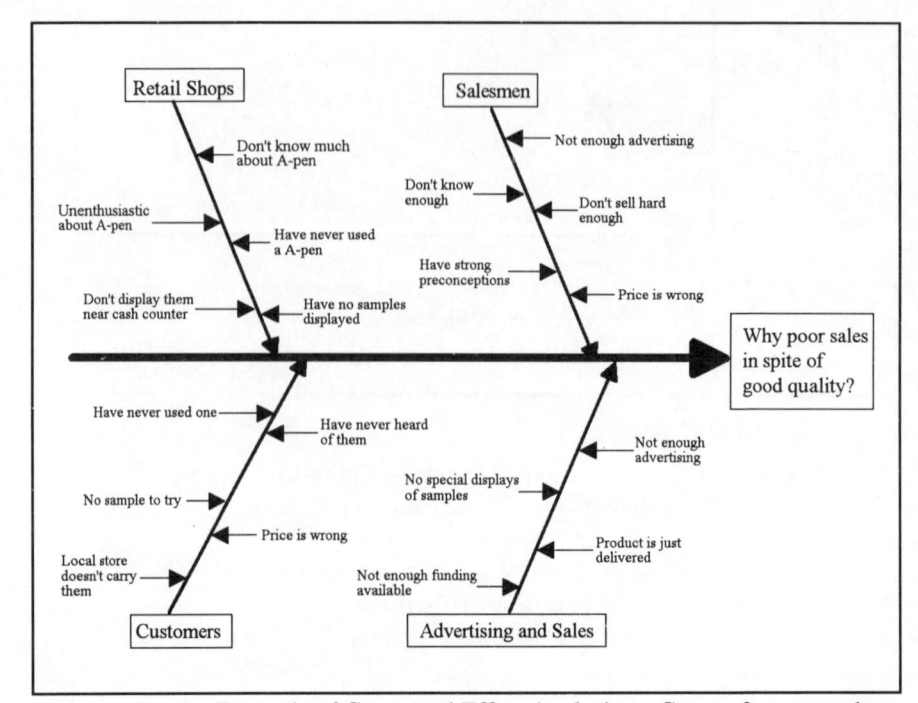

Figure 6.6 An Example of Cause and Effect Analysis -- Causes for poor sales

TOOL #6: *Scatter Diagram*

Scatter Diagram (Figure 6.7) is a picture of possible relationships between paired data. The paired data consist of an independent variable and a dependent variable. For example, the extension of a rubber band is dependent on the force applied at the ends. In this case the extension is the dependent variable and the force the independent variable. If dependence of one variable on another can be shown as a simple graph plot, it may be possible to control the 'dependent' variable by adjusting the 'independent'. Analysing this graph can help to confirm or reject hypothesis about possible correlation between the two variables.

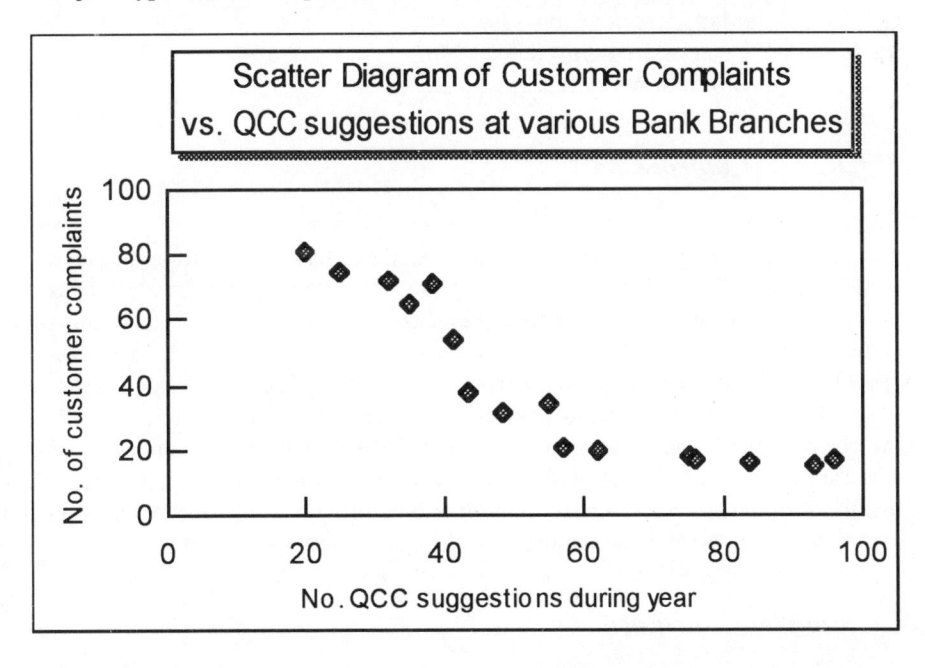

Figure 6.7 An Example of a Scatter Diagram -- Number of Customer Complaints plotted against the Number of suggestions from QCCs of Bank Branches

An important statistical factor to analyse the correlation between the two variables is called Correlation Coefficient. For continuous data (such as extension of a rubber band), Pearson's Correlation Coefficient is a good measure. For discrete data (such as number of defects or customer complaints), Spearman's Correlation Coefficient is widely used. The calculation of both coefficients can be found in most statistical textbooks [e.g., Rees, 1994]. An easy way is to use a statistical software package to work them out automatically. Apart from being able to arrive at the solution much quicker, a good statistical regression package will enable you to identify the *best fit* curve to a set of scattered data based on the highest correlation coefficient from a number of curves equations (both linear and non-linear).

These two coefficients are important because both of them end up as a single coefficient **R** between -1 to 1. The interpretations are:

R	Meaning
-1	Best negative correlation (Negative means when one increases, the other decreases proportionally).
-1<R<-0.7	Good negative correlation.
-0.7<R<0	Bad negative correlation.
0	Worst correlation, no association between the two variables at all.
0<R<0.7	Bad positive correlation (Positive means when one increases, the other increases proportionally).
0.7<R<1	Good positive correlation.
1	Best positive correlation.

TOOL #7: Control Charts

Control Charts (Figures 6.9--6.12) are pictures of variation over time. Data are plotted on graphs against time, preferably at the **point of operation** where the process is actually carried out. Decisions are then made about the state of the process and whether any action should be taken. The UCL and LCL stand for the Upper and Lower Control Limits respectively. They indicate what the process is capable of doing. The variation around the mean is expected and comes from what Deming called 'common' causes. The spikes above the UCL or LCL indicated 'special' causes of variation and can be corrected. The reduction in variation comes from making fundamental changes in methods, machines, or materials. Control charts help to monitor and control quality by acting as a set of process traffic lights, and are valuable in all types of activity.

Control Chart for Variables

Control charts for variables consist of Mean-and-Range Charts. For example, the mean-and-range charts may be used to control the quality of sugar packaging by measuring the weights in random samples, to improve the customer services by measuring the queuing time before they are being served, etc.

Mean Chart

Mean Chart for variables make use of the properties of samples. In practice a continuing series of samples is taken, their means are calculated and each mean plotted on a *mean chart*. Statistically, the mean chart can be expressed as a function of the *mean of all the sample ranges* (W) multiplied by a factor (A) dependent on the sample size as shown in Figure 6.8.

For Mean Chart, Upper Control Limit (UCL) = (Mean of means) + A * W
Lower Control Limit (LCL) = (Mean of means) - A * W

Chapter 6: QCC & Problem Solving

Range Chart

A process may maintain its mean value while the *spread* of the performance may increase. In other words, the accuracy may be unchanged while the precision deteriorates. To monitor this, the ranges of samples are plotted on a *Range Chart*. Just like the mean chart, the range chart can be expressed as a function of the mean of all the sample ranges (W) multiplied by a factor (D) dependent on the sample size as shown in Figure 6.8.

For Range Chart, Upper Control Limit (UCL) = D * W

The lower control limit for range chart is not significant, since a reduction in range implies an increase in precision, and therefore there is no need to control.

NOTE: *A process cannot be considered to be in control unless both the mean and the range values are inside their control lines. Because of this requirement, Mean-and-Range Charts are used simultaneously.*

Sample size (n)	MEAN CHART Factor (+/- A)	RANGE CHART Factor (D)
1	2.73 *	N.A.
2	1.94	4.12
3	1.05	2.98
4	0.75	2.57
5	0.59	2.34
6	0.50	2.21
7	0.43	2.11
8	0.38	2.04
9	0.35	1.99
10	0.32	1.93

* In this special case, the range is defined as the **moving range** which is simply the gap between consecutive observations in the data.

Figure 6.8 Mean and Range Chart Factors for different Sample Size (at 99.9% confidence, or 3.08 standard deviations)

Worked Example 6.1

A machine packs sugar at a rate of 1000 packets per hour. Every hour a random sample of 5 packets is taken and their weights are measured. After 10 hours the data are collected as shown in the following table. Use these data to design control charts for the sample mean and range of the weight concerned.

Sample	Measurements (kg)					Mean	Range
1	1.02	1.01	1.03	1.04	1.05	1.03	0.04
2	0.98	0.97	0.99	1.00	1.01	0.99	0.04
3	1.01	1.02	1.04	1.04	1.03	1.03	0.03
4	1.04	1.02	1.03	1.03	1.02	1.03	0.02
5	0.97	0.95	0.98	0.99	1.01	0.98	0.06
6	1.02	1.03	1.01	1.05	1.04	1.03	0.04
7	1.03	1.06	1.07	1.05	1.01	1.04	0.06
8	0.98	0.99	0.96	0.99	1.02	0.99	0.06
9	0.98	1.01	1.03	0.97	0.99	1.00	0.06
10	0.99	1.03	1.04	1.04	1.03	1.03	0.05
					Mean of	1.012	0.046

The sampling exercise was then continued for the next 12 hours, the data collected are shown in the following table. Comment on the variation of the 12 hours.

Sample number	1	2	3	4	5	6	7	8	9	10	11	12
Mean of sample	1.02	1.03	1.02	1.01	1.02	0.99	0.99	1.02	1.05	0.98	1.03	1.02
Range of sample	0.05	0.07	0.11	0.08	0.03	0.13	0.09	0.06	0.10	0.09	0.04	0.05

For Mean Chart, Upper Control Limit (UCL) = (Mean of means) + A * W
= 1.012 + 0.59 * 0.046
= 1.039

Lower Control Limit (LCL) = (Mean of means) - A * W
= 1.012 - 0.59 * 0.046
= 0.985

For Range Chart, Upper Control Limit (UCL) = D * W
= 2.34 * 0.046
= 0.108

From the Mean Chart (Figure 6.9), Samples No. 9 & 10 are outside the control limits and require action. From the Range Chart (Figure 6.10), Samples No. 3 & 6 are outside the UCL and require action. All these four samples should require corrective and preventive actions immediately.

Chapter 6: QCC & Problem Solving

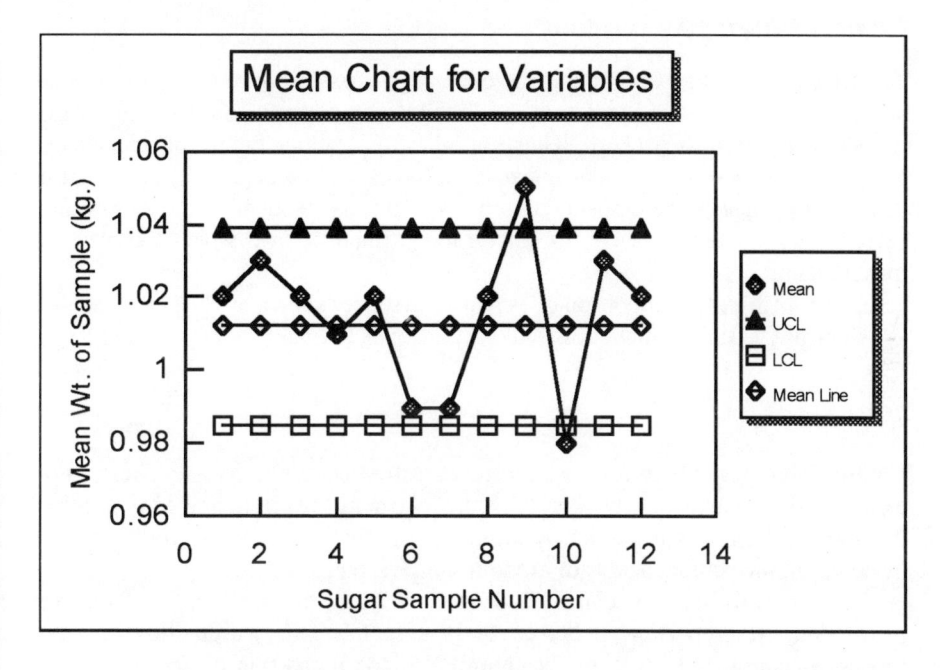

Figure 6.9 Mean Chart for Variables for Worked Example 6.1

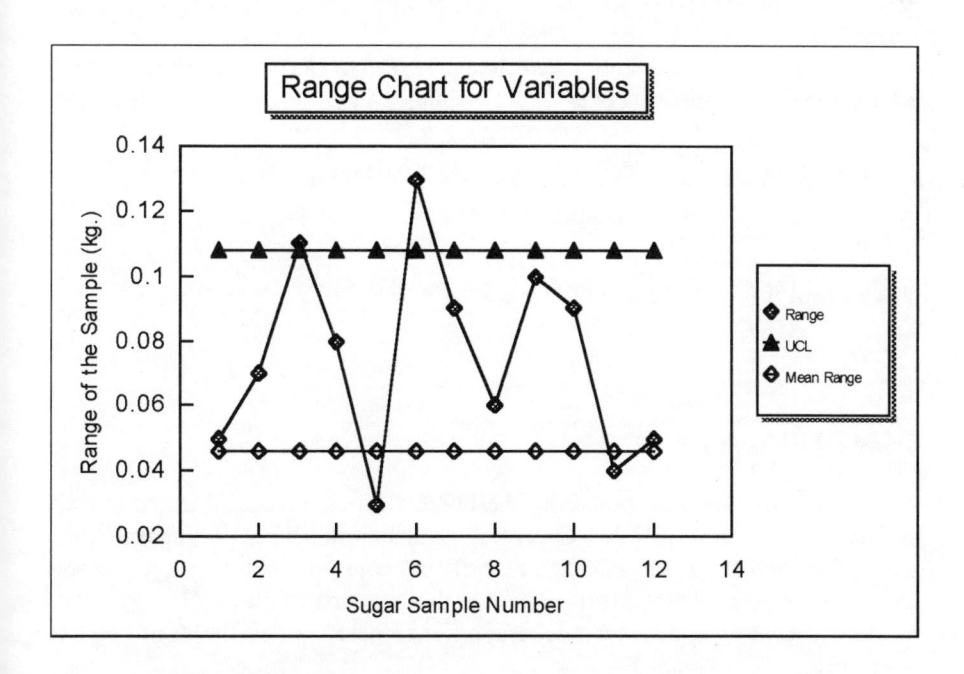

Figure 6.10 Range Chart for Variables for Worked Example 6.1

Control Charts for Attributes

The Mean-and-Range Charts and the example discussed so far are applicable to *variables or continuous data* (i.e. data which varies continuously, such as the weight of sugar in a packet). However, there is another family of data called *attributes or discrete data*. For instance, in the case of acceptance sampling, it is possible after inspecting items to classify them only as 'good' or 'bad'. As a result, there is another family of control charts for attributes. Two types of charts are most popular:
1. Control Chart for Proportion or Percent Defective (known as *P-Chart*).
2. Control Chart for Number of Defects (known as *C-Chart*).

P-Chart

Control chart for *proportion or percent defective* is known as a P-Chart. For example, the P-Chart may be used to control the quality of light bulbs by counting the percent defectives in random samples, to test the punctuality of train services by counting the percent delays in random samples, etc.

The control limit is constructed after a pilot investigation. If during production the proportion of defectives in a sample falls within this limit the process is considered to be 'under control'. If the proportion of defectives in a sample falls beyond this limit, this is taken to be a good indication that the process is, for some reason, out of control and that some investigation and corrective action are required.

An estimate of the proportion defective produced by the process (P) is obtained after a pilot study consisting of at least 10 samples, i.e.,

$$P = \frac{\text{Total number of defectives in at least 10 samples}}{\text{Total number inspected}}$$

The Control Limit at 99.9% confidence level $= P + 3.09\sqrt{\frac{P(1-P)}{n}}$

where   n = sample size

Worked Example 6.2

A railway operation was measuring the number of late arrival of trains at each station along the railway line. Every day a random sample of 100 arrival times were taken and the number of late arrivals are counted. After 10 days, it was found that there were 80 late arrivals out of the 1000 arrival times. Use these data to design control charts for the proportional defective of the train service concerned.

Chapter 6: QCC & Problem Solving 151

The sampling exercise was then continued for the next 12 days, the data collected are shown in the following table. Comment on the variation of the 12 days.

Day number	1	2	3	4	5	6	7	8	9	10	11	12
Late arrival (%)	9	7	5	6	12	14	16	17	18	15	16	14

The proportion defective recorded during the first 10 days, P = 80/1000 = 0.08

Therefore the Control Limit = $0.08 + 3.09 \sqrt{\frac{.08(0.92)}{100}}$ = 0.164

P-Chart for Attributes

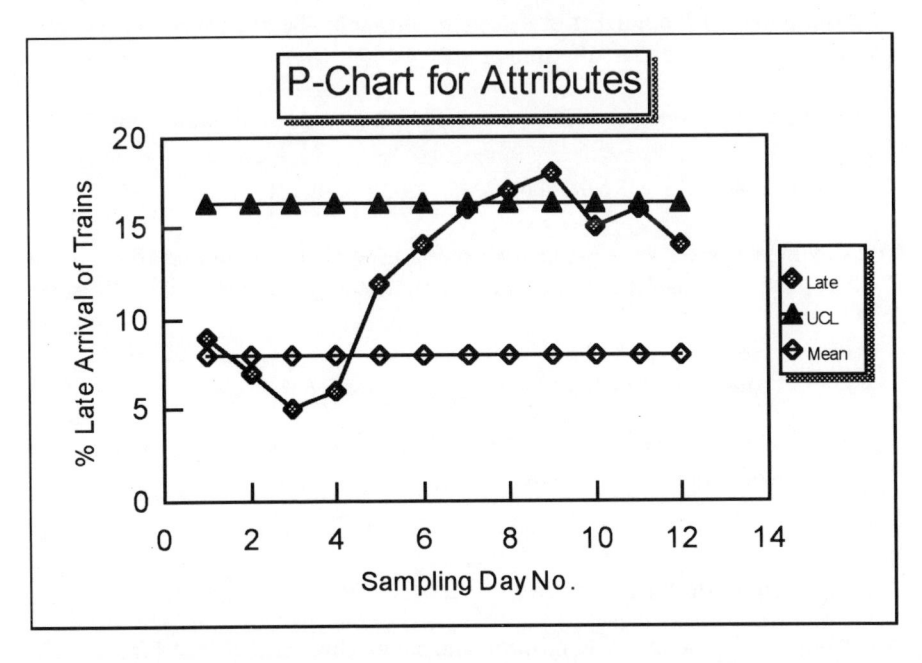

Figure 6.11 P-Chart for Attributes for Worked Example 6.2

The P-Chart is plotted in Figure 6.11. The data for the next 12 days are plotted on the P-Chart. It is seen that starting from Day 8, the data have gone outside the control limit. Therefore, some investigation and action need to be taken. In practice, if the trend goes up, like what has happened during Days 5--7, it is a *warning signal* already, and the reasons for the defective trend should be investigated before it goes too bad. On the other hand, if some improvements have been made to the process (say by better scheduling of the train service), then a new control limit should be used based on the 10 most recent data. Repeating this process will lead to continuous improvement -- *kaizen*.

C-Chart

The control chart for *number of defects* is known as a C-Chart. For example, the C-Chart may be used to control the quality of cloth by counting the number of defects in a roll, to test a piano lesson by counting the number of misplays in a piece of test music, etc.

The control limit is constructed after a pilot investigation. If during production, the number of defects in a sample falls within this limit the process is considered to be *under control*. If the number of defects in a sample falls beyond this limit, this is taken to be a good indication that the process is, for some reason, out of control and that some investigations and corrective actions are required.

An estimate of the number of defects produced by the process (C) is obtained after a pilot study of at least 10 samples, i.e.,

C = The mean number of defects averaged from at least 10 pilot items

The control limit at 99.9% confidence level = $C + 3.09\sqrt{C}$

The way in which C-Chart is constructed and used is very similar to those of the P-Chart. For instance, if the total number of defects in 10 pieces of cloth is 40, then C = 4.

Therefore, the control limit at 99.9% confidence level = 4 + 3.09 * 2 = 10.18

So, in future if the mean number of defects in a sample is greater than 10, then the process will require corrective action.

Example of Process Improvement using Control Charts

An example of the application of mean and range chart is shown in Figure 6.12. It is highlighted that on Aug 17 the process was out of control because the mean reached the UCL of the mean chart. Also on Nov 9, the mean range was also out of control because it reached the UCL of the range chart. Therefore, there is a good reason to suspect that during this period, there was an abnormality. As a result of the investigation, the air valve pressure was reduced on Dec 12. The result of the improvement is that both mean and range charts are under control. In fact the means are very close to the target mean line.

As a result of the improvement in the process, the worker decided to work out the new control limit based on the additional data. It can be seen from the control charts that both the mean and range control limits were reduced on April 12. Then the cycle of improvement had been repeated, and on Sep 22 there was another improvement in the control limits. Now the process is under much tighter control than at the beginning. This example demonstrates that control charts are very useful tools for *kaizen*.

Chapter 6: QCC & Problem Solving 153

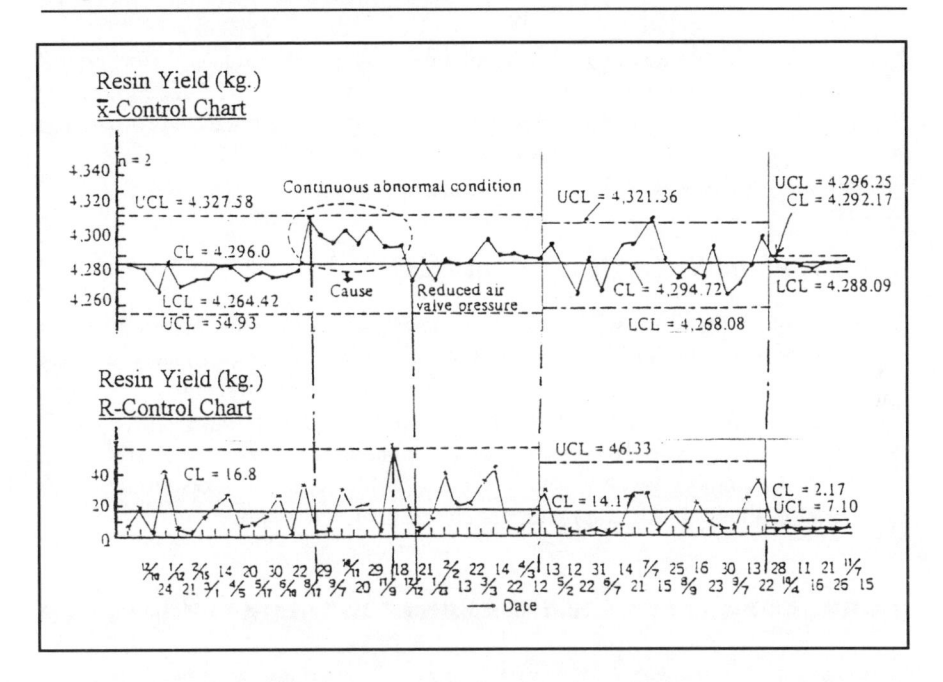

Figure 6.12 Example of Process Improvement using Control Charts --
Resin Yield Force (kg.) [Ishikawa, 1984]

Further Examples on the Use of Control Charts

According to Deming [1993] statistical thinking has been applied to-date to only 3% of business processes, i.e. the factory shop floor. Yet the biggest opportunity for improvement (97%) is throughout the rest of the organisation. **Management can apply the same concepts to any business process involving measurement.** Trying to look for explanations when none exists is one of the mistakes that managers frequently make. A mistaken interpretation of measurements without an understanding of variation (see Deming's *Knowledge of Statistical Theory* in Chapter 2 for details) is more dangerous than no measurement at all. It can make things worse.

Examples of variation that we might come across include [Starkey, 1995]:

On a personal level -- Expenditure on food, the time taken to travel to work, how many miles we drive per year, how much fuel we use in our cars, how much water we consume.

In business -- Annual and monthly sales, profit margins, return on capital employed, production output, delivery times, invoicing errors, number of calls, number of quotations, calls per order, number of complaints.

The economy -- Trade deficit, exchange rates, interest rates, number of new company start ups, number of company failures.

The environment -- Monthly or annual rainfall, sunshine hours, pollen counts, crop yields, fruit size.

More types and applications of control charts can be found in Wheeler and Chambers [1992]; and more examples on the use of control chart can be found in:
Starkey [1995] -- monthly sales revenue, customer service index, product management, sales performance ratios; and
Brewin and Starkey [1995] -- customer claims, on-time delivery.

The Use of the Seven QC Tools in QCCs

The seven QC tools discussed above are necessary for successful performance of a QCC. They form the basis for data collection, discussion and problem analysis. They are important and effective methods for quality improvement. It has been the Japanese experience that 95% of the problems in the workplace can be solved with these seven tools [Ishikawa, 1984]. Therefore, these tools should be promoted and training given to all employees. If the seven QC tools are used by a QCC organisation, quality improvement can be achieved readily.

■■

Mini-Case 6.1:

7-Eleven Using Quality Control Tools as Competitive Advantage [Peppard, 1993]

7-Eleven, with 4,000 convenience stores, is Japan's largest food retailer, and is separately owned from the American company of the same name. Mostly owned by franchise-holders, 7-Eleven has the highest profitability and return on equity among Japanese retailers. The parent, a subsidiary of a Japanese retailer, owns only about 5% of the stores, a figure which is declining. However, 7-Eleven owns a sophisticated computer network which collects sales data and orders goods directly from the distributors.

Each store is equipped with a point-of-sale system. When something is purchased, the point-of-sale records its brand name, the manufacturer, the price, and details about the buyer. Sales of products can then be plotted against the time of day, day of the week. The quality control tools used in the analysis include: Check Sheet, Histogram, Pareto Analysis, Scatter Diagram and Control Charts.

7-Eleven, the parent company, can aggregate data from many stores. It then collates and sells this data to manufacturers about the sales of their own products. These statistics also allow 7-Eleven to advise their franchise holders on what to sell and when -- important considering the average shelf space of only 10 square metres and over 3,500 products. The shelves are filled with a mix of goods according to the time of day.

■■

Problem Solving

"We must acquire the actual experiences of solving problems rather than search for and focus on methods."

-- *Hans J. Bajaria, President, Multiface Inc.*

In nowadays increasingly prevailing TQM environment, problem solving still plays a vital role, and success of businesses relies on good decision making. The main tasks of QCCs are to solve existing problems, predict and prevent potential problems, and institute continuous improvement. The methodology to solve potential problems is very similar to solving existing problems. Continuous improvement is a consequence of the first two tasks. Therefore, solving existing problems is the most important task of QCCs.

Is TQM problem oriented? It is more of a system that, by employing proactive leadership, puts all relevant elements together to create a supportive, learning culture where prevention of failure and success of all people in the supplier-customer chain is an ultimate goal. It should not, and, in reality, does not imply that in a "total quality system" problems will never occur in the course of operation. Working conditions are always subject to ongoing changes in the micro and macro environment, and the organisation must learn how to handle them by utilising its system, structure and skills. Prevention (quality assurance), a step forward in the evolution of total quality, is generated from the experience of problem solving, and results directly from implementation of the Deming's Plan-Do- Study-Act Cycle [Deming, 1993].

Of all the managerial functions relevant to quality leadership, the one attached to managers as "investigators" plays the major role in problem solving. From the announcement of a problem until its resolution, the availability of all relevant information is crucial. Since the same information, in the hands of different managers, might result in distinctly different results, it is evident that successful problem solving involves more than the availability of information. Equally critical is the quality of logic applied to that information.

The common mistake made by managers is that decision making phase often follows straight after the recognition of the actual problem. The proper appreciation of the cause of the problem is often neglected. The vital point in the process of problem solving is this simple rule -- **a problem cannot be solved unless its cause is known.** The S-S Method [Ho & Cicmil, 1995] can be used to plug this loophole.

The S-S Problem Solving Method

One becomes aware of a problem when there is an unwanted outcome or effect of the process or action, and anticipates that a certain event has taken place and initiated the unwanted effect i.e. brought the problem about. Although in practice, it is commonly found that there are always several causes relevant to a particular effect, only one specific event, which in some instances can be a combination of several correlated conditions, represents the real **cause** of the problem [Kepner & Tregoe, 1965]. Therefore, the actual operation of the cause in producing an observed effect is not a matter of probability. It is the understanding and knowledge of this cause-effect relation that a manager can hardly ever be absolutely sure about. It is often difficult to ensure that all the relevant facts about any problem have been obtained.

The **S-S Problem Solving Method** (the S-S Method, for short) decreases this uncertainty by offering a procedure of **defining the real cause among the possible ones**, that, for all practical purposes, would exactly and consistently produce all the facts in **the specific description of the problem**. Problem identification rests upon understanding that the unwanted effect is, in fact, a deviation, discrepancy or imbalance between what **should be** (as planned or as standardised) and what actually **is happening**.

Principles of Problem Solving

> *"One thing at a time."*   -- *Winston Churchill*

This should be a rule for effective and efficient problem solving. It is necessary for management to be aware and fully understand the business situation at any point of time through monitoring and controlling. Whenever there are indications that several problems exist simultaneously, the brainstorming technique should be employed in order to identify them. This will help in prioritising the problems according to their severity. The real problem solving procedure starts once the management have chosen the most significant problem to tackle first. This routine should be considered as the preparatory step to the problem solving.

For instance, consider the situation where the management has realised that the business performance was low in a period of time. They brainstormed it and found that there were several problems related to the low performance: design changes, production problems, high staff turnover, poor sales, low profit margin and excessive bad-debts. In the process of prioritisation, they distinguished the low profit margin as the most significant problem to be solved immediately.

The S-S Method has been developed in an effort to suggest a simple and effective way of getting down to the real cause. The process begins with identifying the most significant problem mentioned above, continues with analysis to find the cause, and concludes with decision making. In the attempt to get rid of

Chapter 6: QCC & Problem Solving

the unwanted effect (problem) efficiently, the real cause must firstly be identified and then removed by appropriate action.

Each of the **constituents** of 'thinking a problem through' has to be precisely defined, since these issues are always sequentially linked and require a systematic approach:
- PROBLEM represents what is wrong -- an unwanted effect that needs correcting or removing,
- CAUSE refers to what brought the problem about, and
- DECISION MAKING encompasses choosing the actions to correct it.

The S-S Method Flowchart

```
Step 1: Process Flow Analysis
            |
            V
Step 2: Problem Definition
            |
            V
Step 3: Identify the Real Cause
            |
            V
Step 4: Decide on & Implement Corrective Action
```

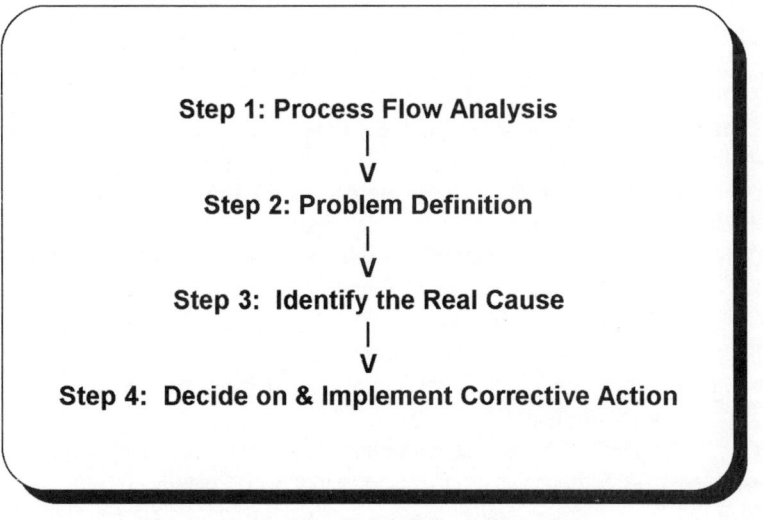

Figure 6.13 The S-S Method Flowchart

The basic structure of the S-S Method has been described in a flowchart, Figure 6.13. Its details are explained as follows with reference to the S-S Method Map (Figure 6.14) and Worksheet (Figure 6.15):

Step 1: Process Flow Analysis
We must know the exact process flow over time when all is going well (Figure 6.14). There is a defined 'should' path that links the beginning of the observed process in terms of time or space, and its end when the outcome can be assessed according to the expectations, stated objectives, or level of performance. But, what is actually going on along that 'path' often varies from what should be going

on. That variation, or 'change', is what should be analysed after the problem has been recognised and defined as worth of acting upon (see Figure 6.14).

Step 2: Problem Definition
After having recognised the deviation or discrepancy in the final outcome that struck us as important, we must be able to give a precise description of the problem through its dimensions regarding nature, people, time, place and extent:
- Use the first column of the *S-S Method Worksheet (Figure 6.15)* to specify the problem by dimensions (note that it still does not uncover the cause).
- Find only the facts that are distinctive to the specified problem, thus eliminating any common factors between what **Is the Problem** and what **Is as Expected.**

Step 3: Identification of the Real Cause -- During the course of the observed process or operation there must be a point when a certain **"CHANGE"** to the regular process flow ('should' path) has happened, and caused the perceived deviation (Figure 6.14). Successful correction or elimination of the problem is only possible if the change, in a form of the single event or condition, is correctly determined.

Therefore, we must track the process flow backwards to find the change i.e. the distinguishing event or condition that caused the problem, through process of selection by looking among all possible causes for those that are limited to the distinctive area of change found in the specification:
- Only changes found in areas of distinctions can be considered as likely causes. The mechanism of search for the real cause can be explained by the S-S **Method Map** as shown in Figure 6.14.
- By analysing the 'should' path of the process only one specific constellation of events (that are new or different to the regular flow) has to be picked up that could uniquely produce that observed outcome -- problem.
- Test the cause found for a real connection with the specified unwanted effect (problem), and then verify if this possible cause would produce the stated outcome.

Step 4: Implementation of Corrective/Preventive Action
Once the real cause has been precisely defined, the decision on the best corrective action can be made and problem solved. So, we should find the best ways to act upon the cause of a deviation by:
- Defining the objectives of decision making and classifying them in order of importance;
- Defining and assessing the alternative actions against the established objectives;
- Exploring future possible consequences;
- Taking other preventive actions.

Chapter 6: QCC & Problem Solving 159

How to use the S-S Method Map and Worksheet

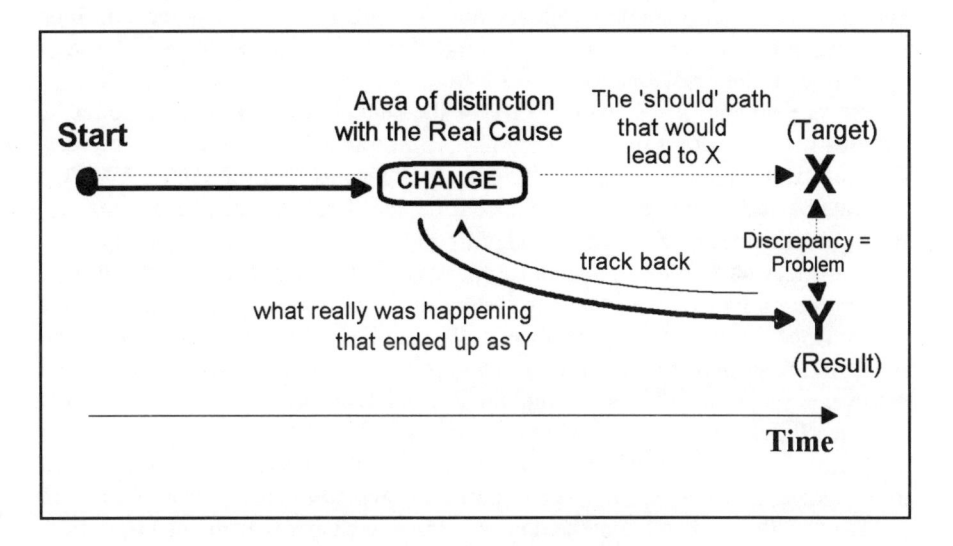

```
Start          Area of distinction    The 'should' path
               with the Real Cause    that would      (Target)
                                      lead to X        X
                       CHANGE                          ▲
                                                       │ Discrepancy =
                              track back               │ Problem
                                                       ▼
          what really was happening                    Y
          that ended up as Y
                                                       (Result)

                                                       Time
```

Figure 6.14 The S-S Method Map

	Is the Problem?	Is as Expected?	The Point of Change
WHAT	Profit reduced by £0.3m	Other performance indicators	Big discount given to a major customer
WHO	(Too early to determine)	The competitors have been doing as we expected	Sales Manager
WHEN	At the end of this year	Last year	6 months before year-end
WHERE	London sales area	Other parts of the country	In London Sales Office
HOW Significant	30% lower than targetted	On target	50% of normal profit margin
Possible Causes that led to the problem	1. Head-on competition with a new rivalry at the time.		
	2. Sales Manager not following pricing policy.		
	3. Poor negotiation tactics with the customer.		
	4. Mistake in calculating cost price.		

Figure 6.15 The S-S Method Worksheet

The S-S Method Map (Figure 6.14) illustrates the problem solving principles and methodology. The S-S Method Worksheet (Figure 6.15) has been designed to help in analysing a problem systematically, defining a distinctive point of change along the pre-determined 'should' path, and narrowing down to the real cause in order to solve the problem and prevent it from recurring.

As an example, let us assume that, at the beginning of the year, a company targeted for a net profit of £1 million (X). Unfortunately, at the end of the year, the profit was only £0.7 million (Y). The difference of £0.3 million (X-Y) is the unwanted effect -- the problem. It should be specified against five dimensions: **what, who, when, where** and **how significant** (the column titled 'Is the Problem' in the worksheet) in order to search for the reason **why** the problem occurred.

By examining other relevant parameters along the 'should' path of the business performance up to date, again using 5-W and 1-H questioning technique and filling in the second column of the worksheet titled 'Is as Expected', management discovered that 6 months ago an unplanned event happened.

The 'Point of Change' column contains data that precisely define this distinctive area to be searched for the real cause of the stated problem - low profit at the year-end. By now, the scope of potential causes has been narrowed down to the issue of the sales six months ago. All possible causes that might have led to this unplanned action are listed in the bottom row of the table. Among these causes, there will be one which coincides with all the symptoms listed in the Worksheet table that could have uniquely produced the observed outcome -- the profit problem. In our fictitious example it could be Cause No. 2, i.e., Sales Manager did not follow the pricing policy.

The last step involves decision making on corrective action and prevention. It might be more centralised control, better communication system and the like.

Case Study I: The 1990 World Cup Semi-Final

The 1990 World Cup Semi-Final between England and West Germany was one of the most exciting matches in that year's World Cup which took place in Italy. The England team came to a 1-1 draw with the West Germany team after 120 minutes of exciting and tough competition. Then the match came to the penalty shoot-out. The results are summarised in Figure 6.16.

The rule of the World Cup Semi-final and final matches is that when it comes to a draw, the winner has to be decided by a penalty shoot-out. Therefore, teams should be prepared to master the situation when it comes up. In football, there are some rules that every experienced football player will agree with:

Rule #1: In a successful penalty shooting, the ball ends up in the goal away from the goal-keeper's reach. The most likely positions are those along the inside edges of the goal-posts, the higher the better, provided that the ball does not go over the bar. The football player must target these points.

Chapter 6: QCC & Problem Solving

Rule #2: A right-handed player is expected to use his right leg for penalty shooting because, most likely, he would have better control of and power in his right leg than the left leg, and vice versa.

Theoretically, in such an important match as the World Cup, the above rules must be adhered to during penalty shoot-out without recourse. This is possible because there is a definite starting point (i.e. 12 yards midway from the goal) and there are no other people interfering, apart from the goal-keeper. Moreover, the football rules favour the shooter because the goal-keeper is not allowed to make any move before the shooter touches the ball. This leads to Rule #3.

Rule #3: The shooter should assume that there is nobody at all in the field, and concentrate on shooting the ball into the positions defined as the best.

ENGLAND			WEST GERMANY		
Goal	Player	Result & Analysis	Goal	Player	Result
E1	Lineker	In	W1	Brehme	In
E2	Beardsley	In	W2	Matthaeus	In
E3	Platt	In -- despite being touched by the goal-keeper	W3	Riedle	In
E4	Pearce	Ball caught by the goal-keeper, player used the left leg	W4	Thon	In
E5	Waddle	Ball flew above the goal, player used the left leg	**WON**		

Figure 6.16 Score Table of Penalty Shoot-out --
World Cup Semi-final 1990: England vs. West Germany

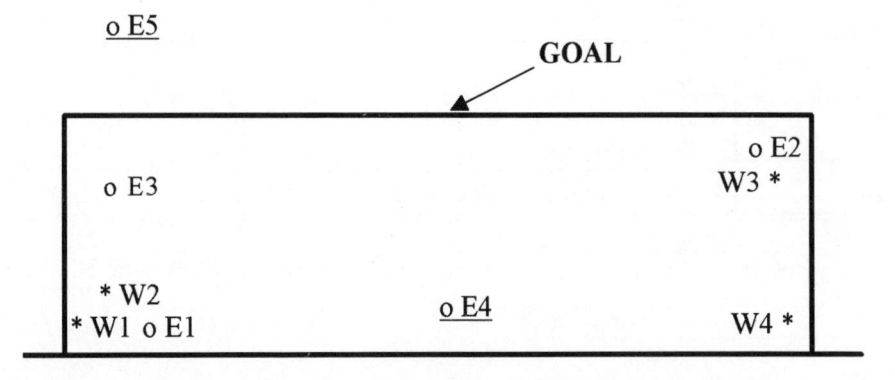

```
o E5
                          GOAL
                                      o E2
    o E3                              W3 *

    * W2
    * W1  o E1         o E4           W4 *
```

Figure 6.17 Approximate Positions of Penalty Shoot-out Goals --
World Cup Semi-final 1990: England (E) vs. West Germany (W)
(Underlined balls indicate missing shoots)

Figure 6.17 shows the approximate positions of the nine penalty goals. Let us try to apply the principles of the S-S Method in analysing such a low performance of the English team players. In this case, the missing shoots are unwanted effects, i.e., problems.

Step 1: Process Flow Analysis -- All players should have followed the Rules #1, #2 and #3 (the 'should' path) without recourse because this would have given the highest chance to get the ball into the goal.

Step 2: Problem Definition -- Figure 6.18 shows the problem analysis of the situation.

	Is the Problem?	Is as Expected?	Point of Change
WHAT	Weak penalty shooting	Performance during the match	Difference in penalty shooting tactics
WHO	2 out of the 5 England players	German players	The way some players shot the ball
WHEN	After 120 min. of match	During the match	Penalty shoot-out took place after a long and tiring match
WHERE	-- At the points easily reached by the goal-keeper -- Above the bar	At the positions near the posts, inside the goal (E2 & E3)	-- Ball easily caught by the goal-keeper or ended up above the bar
HOW Significant	2 failures out of 5 attempts	The German team made no failures out of 4 attempts	Very significant
Possible Causes that led to the problem	1. Some players not following Rule #1 and/or Rule #2		
	2. Certain players are incapable of shooting the ball at the right spot.		
	3. Some players are affected psychologically by the presence of the goal-keeper and have forgotten about Rule #3		
	4. Lack of proper training based on Rules #1, 2 & 3		

Figure 6.18 Problem Solving Worksheet applied to analyse World Cup match problem

Step 3: Identification of the Real Cause -- Although the players are expected to act strictly according to the three rules (the 'should' path) when performing penalty shooting, the area of distinctive change where the real cause of a failure lies is often human psychology. The shooter usually makes guesses on what has been done before him, and what would be the goal-keeper's next guess. This disturbance could affect the decision of the shooter. It is usually at this critical moment that he makes a mistake -- by doing something which is not part of his plan or simply forgetting his original plan completely. Then in most cases, the results are: either giving chance for the goal-keeper to catch the ball (because of the fear of making incorrect guesses) or shooting the ball outside the goal (because of the worry that the goal-keeper might reach the ball).

As the result of the search based on the idea of the Problem Map, **the real cause should be** the lack of proper training. In order to ensure that this is the real cause, we should test it against the What, Who, Where, When and How Significant is the problem:

- WHAT -- Lack of proper training led to the weak penalty shooting, mostly due to players not adhering to Rule #1 and Rule #2 (see Figure 6.16).
- WHO -- A significant number of players were making the mistake as a result of insufficient training.
- WHEN -- When players are tired, the physical condition may affect their decision making. This is why training is important.
- WHERE -- More stringent training on correct shooting (Rule #1).
- HOW SIGNIFICANT -- The importance of the match makes the problem very significant. Therefore training must be thorough.

So, the result was -- England out and West Germany in the World Cup final. In the 1990 World Cup final match, West Germany's opponent, Argentina entirely relied on their superstar player, Diego Maradona. Unlike the 1986 World Cup, when Maradona was well supported with strong wing-players, this time he was left almost alone to make his goal. This 'change' can be identified as the main cause of failure for Argentina, who lost 0 to 1. The assertion made here is that if England had won her semi-final, she could have made her way to the 1990 World Cup Champion, just what West Germany did.

Step 4: Implementation of Corrective/Preventive Action -- The following guidelines should be considered:
- There must be **adequate training** conducted in accordance with Rules #1, 2 & 3.
- Players should be convinced that there are no better alternatives.
- The possibility of penalty shoot-out for future matches should be analysed.
- Preventive actions should be taken to fully understand the psychological effect due to the presence of the goal-keeper.

From this analysis, the Coach must train the players so that they are at the peak of the performance. One very important responsibility of the coach is to train his team for the World Cup again and again on Rules #1, 2 & 3, by putting a dummy goal-keeper at the centre of the goal. This sounds simple but it does work!

Case Study II: The 1994 World Cup Final

During the 1994 World Cup Final which took place in the USA, Italy almost repeated the same mistakes England team had made in 1990. The Italy team came to a 0-0 draw with Brazil after 120 minutes of exhaustive competition. Then the match came to the penalty shoot-out. The results are summarised in Figure 6.19. Figure 6.20 shows the approximate positions of the nine penalty goals.

ITALY			BRAZIL		
Goal	Player	Result	Goal	Player	Result
I1	Baresi	Ball flew above goal	B1	Santos	Ball caught -- too low and not far enough
I2	Albertini	In	B2	Romario	In
I3	Evani	In	B3	Branco	In
I4	Massaro	Ball caught by the goal-keeper -- too low and not far enough	B4	Dunga	In
I5	Baggio	Ball flew above goal			**WON**

Figure 6.19 Score Table of Penalty Shoot-out -- World Cup Final 1994: Italy vs. Brazil

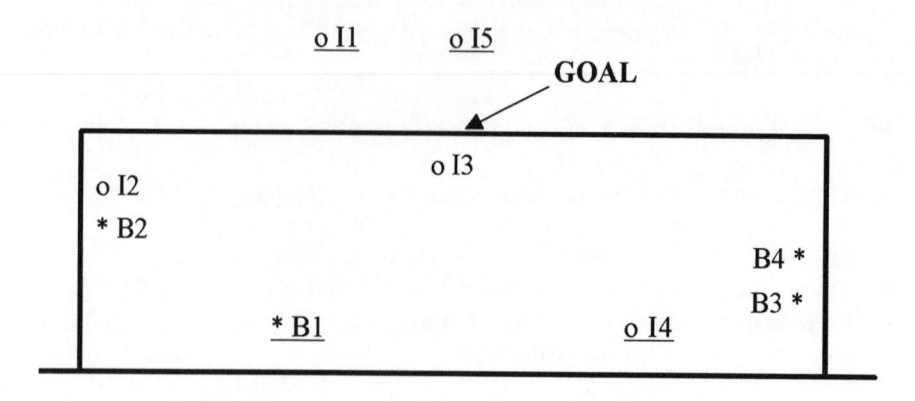

Figure 6.20 Approximate Positions of Penalty Shoot-out Goals -- World Cup Final 1994: Italy (I) vs. Brazil (B)
(Underlined balls indicate missing shoots)

Chapter 6: QCC & Problem Solving   165

On the other hand, as shown in Figure 6.19, Brazil missed the first penalty due to having disobeyed Rule #1, but other team members quickly realised the cause of the failure, implemented corrective action and gave no chance for the mistake to recur. The difference between a winning team and a defeated team is that the winning team (Brazil) could discover the cause quickly and move back to the planned course of action immediately. This difference means success, and is a result of proper training.

Lessons from the World Cup experience

"We have to learn from the mistakes that others make. We can't live long enough to make them all ourselves."
-- W. Edwards Deming

Although the case studies propose how to win the World Cup by using the S-S Method, an even more important objective for managers is to learn from the lesson of the discussed World Cup experience. These lessons can be summarised as follows:
- The cost resulting from a problem can be enormous.
- We must acquire the experience of solving problems.
- The S-S Method is an effective method of problem solving.
- Problems can be solved readily after the real cause is identified.
- We must work towards a plan that prevents problems from happening again.

The S-S Method and the Seven QC Tools

Figure 6.21 suggests how the seven Quality Control (QC) Tools [Ishikawa, 1986] can be used in the problem solving procedure. It is apparent that some of them are supplementary to Steps 1, 2 and 4 of the S-S Method. The table, however, displays that there is no appropriate QC Tool that can replace Step 3 effectively.

In the process of problem solving, identifying the real cause in many cases represents the biggest difficulty. Many people will argue that experience counts, but the fact is that many problems are unique and they may arise for the first time. To be able to prevent the problem, i.e., to ensure that the unwanted effect will not occur again, management must thoroughly understand the nature of the problem, its cause and corrective action required.

| S-S Problem Solving Method | The Seven Quality Control Tools |||||||
	Process Flowchart	Check Sheet	Graphs	Pareto Diagram	Fishbone Diagram	Scatter Diagram	Control Charts
Process Flow Analysis	•	•	•				•
Problem Definition		•	•	•	•	•	•
Identify Real Cause							
Corrective Action	•						•

Figure 6.21 Applications of the 7-QC Tools to the S-S Method

Mini-Case 6.2:

Johnson Electronic Co. Ltd. of Hong Kong [Ho, 1993]

The manufacturer had a plant in Sian, China. There were two in-line ovens. One of them was used to preheat the aluminium circuit board assemblies for powder epoxy coating. Curdling was carried out later in another oven.

One day, the engineer in Hong Kong received a fax from the plant manager in China. He asked for help in the case of three explosions during the injection of powder epoxy. Although no one was hurt, they had seriously disturbed the production process and led to low quality of products. The engineer was experienced in the production line as he was looking after similar lines in Hong Kong. Moreover, he had just finished commissioning a new automatic production line to replace the old one.

He used the S-S Method of problem solving to establish 8 hypotheses. Finally, the evidence confirmed that it was the power failure before the explosions that caused the ventilation system to break down. Thus, the oven temperature was raised too high and caused the explosion of the epoxy. The decision process suggested two measures to be taken:
1. *Ask for schedule of electricity stoppage.*
2. *Replace the conventional system by a fully automated system.*

Then the process continued with potential problem analysis, and implementation and corrective action. Through the rational process, the company managed to install the automatic system in their China plant with very high confidence.

Mini-Case 6.3:

A medium-sized plastic injection moulding factory

The factory had five plunger-operated injection moulding machines with a rigid structure. Each of them carried a mould locking mechanism at one end and a heated cylinder for softening the thermo-setting plastic material at the other. One day, an operator of Machine B found that there were some sink marks and voids on each product produced. The engineer came in to investigate the cause.

The first thing he did was to compare the readings of all machines. He run the Machine B and was surprised to find that the cylinder temperature was much lower than with other machines. After investigation, it was found that the thermocouple was malfunctioning. So he replaced the old thermocouple with a new one and ran the machine again. The extent of sink marks and voids were greatly reduced but still appeared. So he thought that there was still something wrong in Machine B. He finally found another distinctive factor -- the mould temperature was lower in Machine B than in all other machines. Out of the several alternatives, it was confirmed to be the dirt in cooling channels. After cleaning, the machine started to produce good quality parts again.

To prevent future problems from arising again, the engineer suggested that there should be a team of fitters responsible for the maintenance of all the machines. Special care was taken that no dirt, dust and granules were to be entered into the injection moulding machines. Further, there should be daily check on hydraulic oil levels, lubrication, leaks on pressure lines, hose connections, and electrical wires, including thermocouple leads. Every week the machines were to be cleaned thoroughly and the oil checked.

Finally, the management have ensured smooth implementation of the newly installed equipment.

Why do we need QCCs?

Programmes which are based on QCC practices have been introduced for a variety of reasons, but firms invariably find that *the quality of product and service is improved as a result of QCC activities.* QCCs reveal all sorts of faults that prevent good practices, thus improving job satisfaction and contributing to pride in workmanship. This leads to higher quality of products, increased awareness of quality, and continuous improvement.

Another benefit is an improved two-way communication. The management becomes more concerned with the staff problems and, in turn, the staff becomes aware of the day-to-day problems of running an organisation. Communication between departments also improves. While QCCs work on their own area's problems, their systematic approach often reveals previously unsuspected causes of difficulties in related processes of the production flow. A QCC programme in general requires the same framework as ISO 9000 quality standards regarding the management structure and in-company training. Therefore, QCCs should be part of any company's Total Quality Programme.

Everyone's commitment to improvement imposed by a QCC programme also helps to establish customer confidence. Although some companies do not set out to achieve a pure financial return, most find that the financial benefits considerably overrun the costs. Some have experienced ten-fold savings, taking into consideration the gains cumulated year after year.

■■■

Mini-Case 6.4:

Hydrapower Dynamics Ltd. of Birmingham   [DTI, 1992, Mar]

Hydrapower Dynamics of Birmingham was established in 1983, and is an expanding company producing and servicing hydraulic hoses and fittings for use in industry, agriculture, construction, shipping, aerospace and other areas of application. With 18 employees, the company was revitalised by a management buyout in 1989, and began a TQM programme in October 1990, instigated by the Managing Director who led the buyout. His rationale was that Hydrapower Dynamics had attained ISO 9002 and that a Total Quality Programme was the next logical step to ensuring premium performance and satisfied staff and customers.

The first step was to brief all staff on the concepts of TQM and QCC. A seminar was arranged for employees and key suppliers at which a presentation was made by the company's Quality Manager, together with the National Society for Quality through Teamwork (NSQT). The Managing Director then approached the Birmingham Economic Development Unit for advice, and was put in touch with one of their QCC consultants. The Unit sponsored 100 hours of the consultant's time to be devoted to designing and initiating a QCC programme. Under the consultant's guidance, all staff members undertook training in carrying out QCC activities. The training consisted of video material, brainstorming and problem solving sessions. As suggestions were worked on, the advantages of a TQM programme soon became clear to production staff and management, and many of the changes that have been adopted sprang from shopfloor suggestions for improvement.

Hydrapower were not able to organise normal QCCs because of the disruption it would have caused. With just 18 people it didn't make sense to spend working hours discussing ways and means of improving production and delivery because these activities, being unattended in that period, would inevitably be slowing down. *QCC meetings still took place on a voluntary basis after the official working hours. The company formed 'Action Teams', who could rely on their experience for instant problem solving without having to first consult other departments.*

The Managing Director is enthusiastic about the results: *'From a management point of view they have been tremendous, totally vitalising. Management are now more aware of the need to involve staff in decision making, and our staff are far more participatory and enthusiastic about their work and the company as a whole.'* Since QCCs were formed, there have been substantial improvements in all areas of operation. New benches were designed to their own specifications by the workers using them, new test safety features have been incorporated into the testing of high pressure hoses and a new telephone system has been installed to the company's specification. QCCs have researched the time factor between ordering and delivery, and their resulting recommendations have been implemented with very satisfactory results.

Through the TQM programme improvements have been made in customer services with Hydrapower's trade counter excelling in meeting customers' instant needs. Perhaps the best indicator of Hydrapower's achievement in realising its quality goals is reflected in the BSI Inspectorate. Unable to find enough discrepancies, they have reduced their audits by 50 per cent.

■■■

How to Establish QCCs

Companies with the most successful QCC programmes *have spent time in the early stages making sure that everyone in the company is properly informed and consulted before any QCC activity begins.* Often an outside specialist will have assisted with the first awareness presentations. Once established, a typical programme will have QCCs operating in all parts of the company -- in offices, service operations and manufacturing. Experience shows that the size of a company is not important to a programme's success but it significantly affects the support structure and organisation. The steps of implementation are:
1. Management is made aware of the QCC process through a management briefing.
2. The feasibility of the QCCs are analysed.
3. A steering committee is formed.
4. Co-ordinator and in-house instructor is selected.
5. Potential area for initial circles is selected.
6. QCC presentations are made to first-line supervisors in identified areas, divisions or departments.
7. Co-ordinators and middle management receive extensive training on the process and their roles.

8. Supervisors who are interested volunteer and receive training.
9. Following training, QCC presentations are made to the employees who report to the newly trained supervisors.
10. Employees volunteer to be members of a circle and receive training.
11. A circle is formed and begins work.
12. Additional circles are formed as interest broadens.
13. Circles work in a systematic way in solving problems, not just discussing them.
14. Management must ensure that solutions achieve a quick implementation once they have been accepted.
15. Circles are not paid directly for their solutions, but management must ensure appropriate and proper recognition.

Elements of a Successful QCC Programme

"There are three kinds of people:
People who MAKE things happen,
People who WATCH things happen, and
People who WONDER 'What happened?'"

-- *Anon.*

In order to implement QCC successfully, the following guidelines have to be considered:
- Participation is voluntary.
- Management is supportive.
- Employee empowerment is required.
- Training is integral part of programme.
- Members work as a team.
- Members solve problems not just identify them.

QCC Nominal Group Technique

This is a technique for increasing contributions from individuals in a group setting. It is designed to overcome social and interpersonal barriers between people from different levels, social status, or competencies involved in solving a common problem. It is structured so that people generate a list of solutions to a problem individually. For example, a question might be "What are the limiting factors to this company delivering a product on time?" Each participant would write his own list of limitations. All the lists of limitations are collected and made public without comment and criticism. This exercise is completed before anyone talks, thus eliminating any inhibiting factors to influence the problem solving process. The process shares this characteristic with brainstorming. After a period of discussion to clarify limitations and to omit duplications, a vote is taken to prioritise the limitations left on the list. The priority list becomes the basis for

further problem solving. During the brainstorming session the leader should consider the following questions:
- Is everyone thinking about the same problem?
- Are all ideas (good and bad) encouraged?
- Are all ideas recorded?
- Do all members have equal chance to participate?

QCC Code of Conduct

In general, the following code of conduct for QCC discussion applies:
- Criticise ideas, not persons.
- The only stupid question is the one that is not asked.
- Everyone in the team is responsible for team progress.
- Be open to the ideas of others.
- Pay, terms of employment and other negotiable items are excluded.

QCCs and Employee Empowerment

"The Business exists as much to provide meaningful work to the person as it exists to provide a product or service to the customer."

-- Robert Greenleaf

Management needs to address issues concerning people empowerment. The fundamentals of the whole theory of TQM are human resource oriented, emphasising that, in order to make the most of people, management must ensure that:
- employees clearly understand what is required of them,
- all the tools needed for the successful performance of the particular work are available and accessible to the employee,
- people know how the company defines success,
- rewards are based on clearly defined performance standards
- people have knowledge and skills to do the job,
- people take part in problem solving at their workplace.

QCC activities should reflect all the above requirements. TQM does imply some amount of discipline into work schemes through strictly defined individual responsibility and precise procedures and job description. At the same time, though, it enhances an individual pride and sense of purpose by doing a quality job, making the best of his knowledge and skills, and sharing experience with others, contributing to the process of learning and continuous improvement.

Closely relying on Herzberg's [1966] analysis of events at work which produce job satisfaction (5 'motivators' -- achievement, recognition, the work itself, responsibility, advancement), TQM particularly stresses the need of giving people real responsibility for their work and satisfaction in getting the job right first time.

■■

Mini-Case 6.5:

Employee Empowerment at Lyondell Petrochemical   [George and Weimerskirch, 1994]

Lyondell Petrochemical is one of those American companies who were granted the title of Malcolm Baldrige National Quality Award (see Chapter 9 for details on Baldrige), its employee involvement being the key factor for their success. It all started with the change in managerial paradigm of being prepared to 'give employees control over their activities, freedom to make important decisions, and responsibility for their actions - forever'. The transition to employee involvement was a painful process in this company because they had had a long tradition of having clear lines drawn between management and the work force. They managed to communicate the idea of 'integration through quality' by training employees to accept responsibility, giving feed-back, and establishing criteria for rewards and recognition.

To be able to accept responsibility, people need to be trained in their new roles, given opportunities to succeed, supported, and encouraged. The problem Lyondell faced along this route was that some people were scared of their own empowerment because they did not want additional responsibility, and quite a few never made it. One of those who did, a laboratory technician says, "If I see something that needs to be done, I do it or find someone who can. That's empowerment to me."

In this company, managers and supervisors are trained in giving on-the-spot and monthly feedback. There is an established code of behaviours against which every employee's annual performance is evaluated and improvements result in merit pay.

■■

QCC and Specifications

QCCs often have to cope with the issue of responsibility which raises particular problems because of the overriding importance of the specification. All work -- design as well as a finished product -- either does or does not conform to its specification. Thus, the notion of conformance is imposed on quality by definition, and may be seen as a demotivator.

Many sceptics argue that TQM suspends creativity by introducing certain rules of compliance and conformity. One must not forget that quality should be a common language in communication with customers, and that, therefore, the organisation must allow for some trade-offs within its processes.

As it is the case with creativity, the notion of 'conformance to standards' imposed by TQM is seen as too rigid and uniform, and as such, a barrier to flexibility, individualism and opinion exchange, especially in down-up direction on the managerial scale. Personal responsibility increases motivation, in general. To ensure this happens in the area of quality under a person's control, it is necessary that everyone clearly understands the purpose of specifications. The role of communication throughout the structure of an organisation is essential. People must be given clear specification of requirements in order to avoid ambiguity and misunderstandings, and their own interpretation of 'fitness for purpose'.

There are, sometimes, unwritten rules within corporate culture that are being followed, leading to a complete failure of communication, because line managers are not aware of the situation. It often happens in factories in which the process of design is entirely divorced from that of manufacturing, following different 'ideas' of, for example, drawing tolerance. (Here comes the need for concurrent engineering into focus, but it is a different issue.)

In pursuing the idea of **zero defects,** and attempting to base the personal responsibility for quality on it, it is ethically correct that companies ensure that employees do have the right tools and equipment good enough to produce work which is always within specifications. If not, then the company must explicitly acknowledge that some discrepancies are expected to occur, and it is the employees' responsibility to separate out the work that does not conform to specification.

In these circumstances a random audit is normally carried out. If perceived as a lack of trust, audit demotivates people. From the 'positive' stand point, though, **the audit** is not to find how much bad work has been produced, but to check that the employee has correctly identified defects in the output. In that context, the audit is accepted as 'trust with occasional verification', with the same approach adopted with managers in 'giving them targets and scoring their performance against those targets'. This type of audit is best to be carried out by the QCC Steering Committee, so that it can been seen as an audit for recognition rather than inspection.

Chapter Summary:

- *QCC is a small team of people usually coming from the same work area who voluntarily meet on a regular basis to identify, investigate, analyse and solve their work-related problems.*
- *The seven quality control tools are: Process Flowchart, Check Sheet, Histogram, Pareto Analysis, Cause & Effect Analysis, Scatter Diagram and Control Charts.*
- *QCC and problem solving -- Problem solving is a process that follows a logical sequence. The process begins with identifying the problem, continues with analysis to find the cause, and concludes with decision making.*
- *The S-S Method of problem solving actually is a set of procedures consisting of process flow analysis, problem definition, identification of the real cause, and implementation of corrective action.*
- *QCC is needed because the quality of product and service is improved as a result of QCC activities.*
- *To implement QCC -- Company must spend time in the early stages making sure that everyone in the company is properly informed and consulted before any QCC activity begins.*
- *QCC needs employee empowerment. Moreover, the specifications should not be set too rigidly for QCC to improve.*

Hands-on Exercise:

1. Apply the seven QC Tools to your job and then your organisation. Do you find that there are some gaps which require your immediate attention?

2. What is the relationship between the seven QC Tools and S-S Method of Problem Solving?

3. Apply the S-S Method of Problem Solving to your job and then your organisation. Do you find that there are some gaps which require your immediate attention?

4. How can you install QCCs in your organisation? What are the important considerations?

5. Are QCC and Problem Solving important steps in implementing TQM? Why or why not?

6. How do you think the empowerment of your subordinates can enhance the performance of QCCs in your company?

Chapter 7

ISO 9000 & Quality Audit

Topics:

- What is ISO 9000?
- The content of ISO 9001
- Quality Audit
- Why is ISO 9000 and quality audit needed?
- How to implement ISO 9000
- Benefits of implementing ISO 9000

```
┌─────┐
│ 5-S │
└──┬──┘
   ↓
┌─────┐
│ BPR │
└──┬──┘
   ↓
┌──────┐
│ QCCs │
└──┬───┘
   ↓
┌─────┐
│ ISO │
└──┬──┘
   ↓
┌─────┐
│ TPM │
└──┬──┘
   ↓
┌─────┐
│ TQM │
└─────┘
```

Introduction: What is ISO 9000?

After the Second World War pressure for quality came from the military. As a result, 05 series of Ministry of Defence (MoD) quality standards and the Allied Quality Assurance Publication (AQAP) series of NATO standards were developed in USA and Europe respectively. Major companies in the automotive industry began to establish their own quality system standards and assess their suppliers. In order to control the increase of different types of quality system standards and to reduce the multiple assessments, British Standards Institution (BSI) eventually developed the military standards into the BS5750 series (Part 1, 2 and 3: 1979).

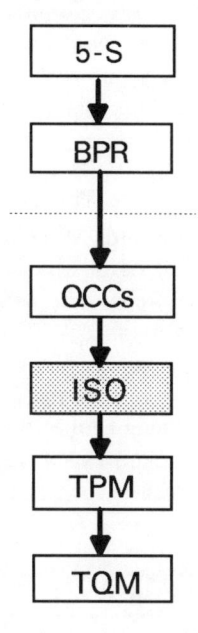

The ISO 9000 series of quality management and assurance standards has been developed by a technical committee (TC 176) working under the International Organisation for Standardisation (ISO) in Geneva. The series was several years in the making before its publication in 1987. In 1987, the ISO adopted BS5750 as a full international standard, the ISO 9000 series, and as a European standard, the EN 29000 series. The second edition came out in 1994.

ISO 9000 Series has already been defined in Chapter 3. The ISO 9000 family comprises of 17 different standards which are listed in **Annex 7.1**. Out of these 17, only the ISO 9001, ISO 9002 and ISO 9003 are **quotable standards**, i.e., can be audited against. In fact, some 99% of the ISO registered firms are registered under ISO 9001 or ISO 9002. In terms of the contents ISO 9002 is a sub-set of ISO 9001. The remaining 14 standards are **guidelines** only. Therefore, for the purpose of this Chapter, only the ISO 9001 will be discussed.

ISO 9001:1994 is the *Quality systems -- Model for quality assurance in design, development, production, installation and servicing*. It is the most comprehensive model of quality systems offered by ISO.

ISO 9001 is significantly different from normal engineering standards such as standards for units of measure, terminology, test methods, product specification, etc. They comprise certain generic characteristics of management practice, usefully standardised, giving mutual benefits to producers and users alike, in a WIN-WIN business relationship. Therefore, ISO 9001 requirements are complementary, not alternative, to the technical (engineering) standards' requirements.

ISO 9001 requirements certainly belong to a TQ process. They cannot take place of a TQ effort because they do not necessarily have to deal with issues of leadership, strategic planning, benchmarking and employee empowerment - which are central to total quality. ISO 9001 does, however, provide a comprehensive approach to documenting quality processes and assessing their performance.

A survey of the total number of global ISO 9000 certifications has been carried out using the resources of the Mobil Oil Corporation. Key findings show that, to the end of June 1994, at least 70,517 ISO 9000 certifications had been achieved in 76 countries - a significant increase from the Jan 1993 figure of 27,824 [Quality World, 1995]. The breakdown of the world share is shown in Figure 7.1.

Region	% Share in Jan 1993	% Share in June 1994
UK	66.8	52.2
Rest of Europe	16.2	26.3
North America	4.3	6.9
Australia/New Zealand	6.7	6.6
Far East	2.4	4.4
Rest of the World	3.6	3.6
Total No. of Firms ==>	27,824	70,517

Figure 7.1 The World Share of ISO 9000 Certifications
(Source: The Mobil Survey, 1994)

The Purpose of ISO 9000 Series

ISO 9000 series sets out the methods that can be implemented in an organisation to assure that the customers' requirements are fully met. Moreover, the organisation's requirements will be met both internally and externally and at an

Chapter 7: ISO 9000 & Quality Audit   177

optimum cost. This is the result of efficient utilisation of the resources available, including material, people and technology. In a nutshell, ISO 9000 is a:
- documented system to interpret the customer's requirements (see Figure 7.2)
- commitment to meeting the customer's requirements
- total company involvement
- nationally accepted standard
- method of cost reduction
- 'springboard' for more effective management control

Figure 7.2   ISO 9000 can minimise misinterpretation of requirements
[Migliore, 1992]

The Content of ISO 9001

The Quality System Requirements (Section 4 which is the most important section) of ISO 9001 has *20 clauses which stipulates the conducts for a good quality management system.* They are listed in **Annex 7.2** for ease of reference. From another perspective, *the 20 clauses of ISO 9001 can be grouped under the following business functions of an organisation.*

Key Elements of a Quality Management System :
- Policy
- Management & Control
- Records
- Audit & Review
- Corrective Action

Meeting Customer Requirements :
- Contract Review
- Design Management Cycle

Control over External Factors :
- Purchasing
- Customer Supplied Materials
- Sub-contractors

Control over Internal Factors :
- Documentation
- Human Resources
- Equipment & Environment
- Instruction & Standards
- Product Identification & Traceability
- Control of Non-conforming Materials

Demonstrating Conformance :
- Inspection & Test Procedures
- Test Tools & Equipment
- Indication of Status
- Use of Statistical Techniques

Preserving Product Quality :
- Handling
- Storage
- Packaging
- Delivery

After-sales :
- Servicing
- Customer Feedback

*After you have a good understanding of the 20 Clauses of ISO 9001:1994, it will be useful to do the **Self-Assessment Exercise 7.1** at the end of this Chapter for your own evaluation.*

Win-Win Situation

ISO 9000 provides assurance to customers that the company can provide the required quality products and services. This can create a win-win situation between the company and its customers which means that both parties benefit from the adoption of the standard. The benefits can be summarised as follows:

Chapter 7: ISO 9000 & Quality Audit

Benefits to the Company
- Increased productivity and efficiency
- Reduced waste
- Reduced time to re-work and re-check by customers
- Improved profitability and performance
- Gained marketing advantages
- Established quality image and trust
- Increased overall business

Benefits to the Customers
- Increased fitness for use
- Increased satisfaction
- Saved time to re-check work done
- Increased confidence and trust
- Saved time to search for and select suppliers
- Reduced cost

An example of win-win situation using ISO 9000 is given in Mini-Case 7.1.

■■

Mini-Case 7.1:

Proposals to Prevent Collapse of High-rises [Ho, 1994, Feb]

When I was working in Malaysia, there was a disaster caused by the lack of maintenance. A 12-storey luxurious residential block in Kuala Lumpur called the Highland Tower (Block A) collapsed on 11 December 1993 and buried over 48 victims. Much has been said about how to avoid similar incident from occurring again. What are the measures from ISO 9000 which can effectively stop further disaster of this kind?

In the building industry, the supply chain starts from the developer. This is followed by architect, contractors, back to developer, and then tenants. So far, there is not yet a developer, architect, and building construction firm in Malaysia which has acquired the ISO 9000 standard.

There is a loophole in the life-cycle of a building. After completion of the building, the contractor has finished his responsibility. So have the architect, and the developer. One may say that there is still a tenants association. Unfortunately, it is most likely concerned about day-to-day activities like cleaning, security, etc. It is unreasonable to expect the tenants association to be competent enough to carry out the detection work on structure and foundation, unless they are properly trained.

In Clause 4.19 of the International Quality Management Standard ISO 9001 or ISO 9002, where servicing is specified in the contract, the supplier (which could be contractor, architect, and developer) has the responsibility to establish

and maintain procedures for performing and verifying that servicing meets the specified requirements. Therefore, the Malaysian Government should enforce the need for a contractual requirement for after-sales servicing for buildings. Then the registered firm should consider the following:
- Clarification of servicing responsibilities among developer, architect, contractor, and tenants;
- Planning of service activities, whether carried out by the supplier, a separate agent, or the tenants;
- Validation of design and function of special-purpose tools or equipment for servicing buildings after installation;
- Control of measuring and test equipment used in field servicing and tests;
- Provision and suitability of documentation, including instructions for use in dealing with the spares or parts lists, and in servicing of the building;
- Provision for adequate back-up, to include technical advice and support, and spares or parts supply;
- Training of servicing personnel;
- Feedback of information that would be useful for improving building or servicing design.

The idea is to build into the ISO 9001/2 quality management system a procedure for regular detection of structural safety. This procedure should be properly documented, with relevant site plans and detail diagrams. Then, in Clause 4.5 (Document Control), there needs to be a clear instruction on the circulation and amendment to the after-sales servicing requirement. Ideally, the tenants should be given a set of work instructions which enables them or the Tenants Association to do regular site inspection, without overburdening them with too much technical details.

ISO 9001/2 is to provide preventive measures to problems and for quality improvement, rather than to correct. Therefore, firstly, the architects need to be registered with ISO 9001. Secondly, when they develop the quality manual for the international standard, they have to consider the significance of Clause 4.19 (Servicing) carefully. They must be able to pass on a simple and clear maintenance plan to the contractor, the developer and the tenants.

The same idea could be applied to the contracting firms. They should also be registered under ISO 9001/2 and carry out the same jurisdiction as the architects have done. The idea could even be applicable to the developers, bearing in mind that they are also part of the supply chain for building construction. The advantage of providing the maintenance plan is obvious: it is a big leap ahead of their competitors who have not done so. They can even advertise such provision to attract more buyers for their properties. Finally, the tenants or the tenants association should take heed of what they are asked to do to protect their own properties, and lives!

Quality Audit

Quality auditing is an important activity that allows departments to continuously improve their function. In the very early stages of ISO 9000 implementation, trained quality auditors/assessors should provide valuable assistance and advice to the Quality Steering Committee (explained later in this Chapter) on how to prepare for registration. Unfortunately, in some cases, quality audits are either not taken seriously or misunderstood. When quality audits are not taken seriously, which is usually a sign of limited managerial commitment, corrective actions remain pending.

Unable or unwilling to correct the system, departments or department heads wrongly assume that it is the responsibility of the quality audit team or the quality manager to address all corrective actions. Quality Auditors should not be perceived as policemen or enforcers of the ISO quality assurance system. Their function is to point out deficiencies or inaccuracies within the system. Moreover, since they are not the line staff of the department being audited, they are not accountable for the remedial actions of the nonconformity they raise. These issues are the direct responsibility of the non-complying department.

Definition of Quality Audit

According to ISO 8402:1994, Quality Audit is *a systematic and independent examination to determine whether quality activities and related results comply with planned arrangements and whether these arrangements are implemented effectively and are suitable to achieve objectives.*

Types of Quality Audit

There are three general classifications of quality audit.

1st Party Audit -- also known as *Internal Quality Audit.* It takes place when an ISO 9000 firm (registered or to be registered) does its regular quality audit internally in order to ensure that the quality management system is implemented effectively. This is a continuous process performed by a few trained people from each department and should be done at least once every six-months. As a normal practice, the auditing function should be carried out by personnel *independent* of the tasks which are being audited.

2nd Party Audit -- also known as Vendor Quality Audit. This is when a firm audits its suppliers who may or may not be a registered firm. Auditing procedure should be in accordance with procedures in the purchaser's quality manual, and they should be communicated to the suppliers well before the commencement of the 2nd party audit. Normally, 2nd party audit is not required if the suppliers are registered firms.

3rd Party Audit -- also known as Adequacy, Compliance or Surveillance Audit. This is carried out by a certified agency, which in the case of the UK, must be registered under the National Council for Certification Award Bodies (NCCAB). If the auditors are coming to the firm to do a mock audit before the actual one, this is called Adequacy Audit. If they are coming to do the actual audit, this is called Compliance Audit. After the firm obtained the ISO 9000 certificate, they still have to come about twice a year to do the Surveillance Audit, in order to ensure that the system is running effectively and to renew the certificate annually.

NOTE: In the subsequent discussion, the principles and practices could be adopted and adapted to both 1st, 2nd and 3rd party audits, though some appropriate modifications and interpretations are required.

How to Facilitate a Quality Audit

1. The Need for Quality Audit

The conduct of quality audits is clearly specified by ISO (paragraph 4.17 of ISO 9001). Moreover, paragraph 4.1.2.2, entitled Verification Resources and Personnel, specifies that "The supplier shall identify in-house verification requirements, provide adequate resources, including the assignment of trained personnel, for management, performance of work and verification activities including internal quality audit". To ensure compliance with paragraphs 4.17 and 4.1.2.2, many companies enrol several employees in one of five day lead assessor courses currently offered by a number of the training organisations approved by the National Accreditation for Certifying Bodies [Dodkins, 1994]. Some larger organisations have even invited consulting agencies for in-house two day auditor training for twenty or more of their employees.

2. Elements of Successful Interviews

Most quality audits are done by interviews. Since an interview is essentially a method of collecting information, steps must be taken to ensure that pertinent information is being gathered. There are several key factors involved in conducting an interview:
- No interview should ever be conducted without a plan which has:
    -- An established objective outlining who within the organisation will have to be interviewed.
    -- A set time-frame to begin and end.
    -- Allocated time for research of the subject to be covered and the functions of the interviewees.
- Keep the interview to the subject for which it was planned.
- Guide the interview with your questions.
- Listen and be sure that you understand what was said and meant. Don't hesitate to cross-check facts with other individuals within the organisation.
- Evaluate the method and the results of the interview.

Chapter 7: ISO 9000 & Quality Audit

3. Preparing for the Interview
Before conducting an audit of your department or internal supplier you should try to obtain the following information:
- An organisation chart. This will allow you to plan your audit as well as estimate how much time your interviews will take.
- A schema of the manufacturing process(es). This can help you formulate specific questions during the audit. Naturally, this step might not be required if you are already familiar with the process.

4. Conducting the Interview
Since most audits can last from 2--5 days, interviewing 8--12 people or even more, it is impossible to remember who said what and when. It is therefore a good practice to have a handbook whenever touring a plant or interviewing people. For each interview, make sure to include time of day and name of the person being interviewed. If people find it difficult to talk to you while you write notes, try to either limit your note taking to keywords, or simply stop writing at all.

Whether you take complete or partial notes, you should always allow ten to fifteen minutes at the end of each interview to prepare your own summary of what has been said.

Effective listening is not only the key element to any good audit, it is also a surprisingly exhausting task. It is therefore recommended that in order to remain sharp and attentive, the auditor should schedule at least two pauses a day.

5. Relation Auditor-Auditee
In order to avoid mistrust and doubt, the auditor should always clearly state the purpose of his/her audit. This step is equally important for quality audits. The purpose should simply be to verify the effectiveness of the quality assurance system. If the system is ineffective, then recommendations must be made on how to improve the system rather than blindly enforcing it. Punitive actions, negative and particularly destructive criticism or sarcasm must be avoided. Such actions can only lead to catastrophic results.

6. Formulating Questions
A good question is a question that:
- Helps the auditee express himself.
- Does not accuse or attack.
- Channel information
- Provoke a reaction
- Provide explanations
- Explore, clarify, verify, illustrate, etc.

The quality audit should be designed to collect facts not vague statements; precision is essential. However, questions should not refer to opinions, feelings or vague concepts.

7. Repeating

It is often very helpful to repeat the answer(s) you have just heard to confirm your understanding with the auditee. This mirror effect can be very helpful. Make use of the following phrases:

"To summarise, if I understand what you just said...."
"Therefore, as far as you are concerned...."
"It is therefore your opinion that...."

8. Some Advice on How to Ask Questions

Since questions can be a very effective tool to acquire information they must be carefully stated. Do not forget that from the point of view of the auditee, your questions will generally be perceived as a form of intrusion. Every time you ask a question you are in some way imposing your frame of reference within which the auditee must try to answer to the best of his ability. Therefore in order to smooth the process, you should:

- Avoid asking questions too rapidly. Do not rush through your interview.
- Avoid asking more than one question at a time.
- Avoid asking lengthy questions.

9. Auditor's Behaviour

An auditor should 'play a role" of a professional. In order to be perceived as one, he should act on the following behaviour without failure.

- He carefully and attentively listens.
- He shows a genuine interest in the explanations provided by the auditee.
- He never judges or criticises.
- He is patient.
- He neither approves nor disapproves.
- He does not prepare the next question while listening to an answer.
- He should never be vindictive, nor suspicious but rather trustworthy.
- He maintains eye contact.

10. Where to Start?

Unless you are interviewing a particular department, your first meeting should always be with the plant director/manager and his staff which would include the quality manager or his delegate. Having explained the purpose of your visit, you should then outline your procedure for the next few days. Normally, you can follow the sequence of the 20 clauses in the ISO 9001 to do your auditing. All along, the guiding principle should be to always ask for documented evidence of the stated (i.e. verbalised) procedures.

11. How Much Time should be Spent?

Individual interviews should generally last between thirty to fifty minutes. When terminating an interview the interviewer should always do three things:

- Give the respondent one last opportunity to mention anything he thinks is significant that hasn't been discussed.

Chapter 7: ISO 9000 & Quality Audit 185

- Get the respondent's permission to return with any questions the interviewer may subsequently think of.
- Get the respondent's commitment to review the interview report.

On average, for a two or three day audit, one can expect to spend four to eight hours interviewing people, a half day checking documentation and between five to eight hours touring the plant. You should leave an hour for the last and most important meeting, i.e., debriefing your host.

12. The Report
The final report should include:
- A list of the persons interviewed.
- A summary of the audit outlining the strong points and the areas needing improvement.
- If appropriate, the report could include pertinent quotations collected by the auditors during his visit. Such quotations can be extremely valuable when making a point.
- If a questionnaire was used during the audit, the scored questionnaire should be returned.
- If the audit is a supplier quality assurance type of audit, the auditor could conclude his report by offering suggestions on how both parties could work together to help resolve problems.
- A date specifying when and by whom the particular corrective action(s) will be resolved must also be set.
- Naturally, records of the audit must also be maintained.
- The audit process itself should be periodically reviewed by management.

13. Presentation of audit report
Finally, the presentation of audit report needs to be done professionally. It will demonstrate the competence and experience of the auditor. Therefore, adhering to a set of rules will help to smooth the reporting process and give chance for the Management Team to response. The suggested procedure for verbal presentation is:

    Team Leader: a) Introduce Team Members
                            b) Scope of Audit (ISO 9001: 1994)
                            c) Area audited
                            d) Disclaimer of the limitation due to observation
                            e) Summary report
                            f) Hand over those reports critical to quality to the Management
                            g) Invite Team Members to read reports
    Team Member: a) Report No.
                            b) Area audited
                            c) Nonconformances
                            d) Requirements of ISO 9001

e) Category - Major or Minor
f) "Do you accept this ?"

Team Leader: Summarise the nonconformance reports that have been agreed upon, and draw a clear conclusion as to what further works have to be done before recommending compliance.

Some Hints for the Auditees

Whenever an organisation or department is about to be audited, a certain amount of preparation needs to take place.

- Make sure you know well in advance (three to four weeks notice is considered acceptable) as to when the audit will take place. Some quality auditors prefer to conduct their audits unannounced, perhaps hoping to catch the villains in a flagrant act of nonconformance.
- If applicable, request any documentation (questionnaire or procedures) relating to the audit procedure. Study the document and make sure you are prepared.
- It is often a good idea to rehearse an audit. In most cases, the pre-audit should be conducted by an outside (independent) source. This will guarantee objectivity. One does not rehearse in the hope of fooling the auditor (although this is often the intent).
- When the auditors arrive, do not try to hide your weaknesses as this will encourage a good auditor to be even more inquisitive. On the other hand do not volunteer information not requested by the auditor. This can be achieved by keeping your answers short and to the point. Make sure you have all pertinent documentation ready for the auditors' inspection.
- Finally try not to crowd the auditors. Although it is always a good idea to escort auditors in order to "control" the audit as much as possible, you should also demonstrate enough confidence in your organisation by allowing auditors to explore wherever they want and interview whomever they want.

*After you have a good understanding of what ISO 9001 Quality Audit is about, it will be useful to do the **Self-Assessment Exercise 7.2** and 7.3 at the end of this Chapter for your own evaluation.*

Why are ISO 9000 and Quality Audit needed?

Irvine [1991] points out that many companies are now seeking registration to quality standard ISO 9000 *to demonstrate that they are in control of their business, and have proved it to a certification body.* ISO 9000 registration is a good way of measuring progress and monitoring maintenance of the standard. It brings marketing benefits, but should be regarded as the beginning of a continuous improvement process rather than the end.

Chapter 7: ISO 9000 & Quality Audit

In many developed and developing countries, ISO 9000 is promoted by government bodies. The UK Department of Trade and Industry has published a full range of booklets and videos on TQM. The Single European Marketing Directive on Standards and Certification recommends ISO 9000 be adopted by all its 16 members. In the South-East Asian countries like Hong Kong, Malaysia and Singapore, governments have set up special divisions to help industries to register for ISO 9000 accreditations. Many Japanese companies in these regions like Sony, Toshiba, Panasonic, etc. have followed. In mid-1992, the Chinese Government required that all foreign manufacturers investing in China and exporting their goods should seek ISO 9000 registration. All these moves prove that ISO 9000 is becoming a necessity in business.

Melville and Murphy [1989] of GEC Plessey Telecommunication Ltd. said their company had chosen ISO 9000 as part of the Total Quality Improvement Programme because they wanted to move away from the traditional role of chasing failure, towards an attitude of prevention where every individual is responsible for producing good quality products and services.

The EC Council Resolution on a global approach to conformity assessment [DTI, 1990] provides three reasons why companies should implement a quality system based on ISO 9000.
- To improve awareness of quality and have the standard for UK products,
- To reduce the need for customer supplier demonstration of quality assurance procedures by introducing third party Quality Assurance certificate,
- To open markets outside the UK by ensuring that ISO 9000 is compatible with EEC and USA quality procedures.

Whittington [1988] in his study to assess the interest for organisations in implementing ISO 9000 and the difficulties they faced, discovered four different reasons for implementing the standard.
- Due to pressure from large customers,
- To maintain contracts with existing customers,
- To use the constraints of the standard to prevent scrap,
- To reduce auditing of the quality system by customers.

Failure to implement the standard for the right reason may prevent companies from gaining the potential benefits from the system. Two of the companies studied by Whittington claimed that ISO 9000 costs much money to implement and maintain, and that their product quality is no better than before the system was implemented. He also found that there was no reduction in assessment and auditing as claimed by much of the literature. Inappropriate reasons for implementing the standard, according to Whittington, are:
- To make reference to the standard on company letter-head paper,
- To get the kitemark symbol on the company's product,
- To enforce discipline on employees,
- To retain existing customers.

Besides the right reasons, the degree of commitment by top management will determine the success of the system. Top management needs to generate a conducive environment to enhance the development of the system. This can be achieved by developing a company quality policy and objectives. This will enable all the employees to work towards the same quality goal.

How to implement ISO 9000

Implementation of ISO 9000 **affects the entire organisation** right from the start. If pursued with total dedication, it results in 'cultural transition' to an atmosphere of continuous improvement. How difficult is the process of implementing ISO 9000? The answer depends on:
- The sophistication of your existing quality programme
- The size of your organisation
- The complexity of your process

There are **9 essential steps** to be followed through in order to implement ISO 9000 successfully.

Step 1:   Top Management Commitment
Step 2:   Establish Implementation Team
Step 3:   Assess Current Quality System Status
Step 4:   Create a Documented Implementation Plan
Step 5:   Provide Training
Step 6:   Create Documentation
Step 7:   Document Control
Step 8:   Monitor Progress
Step 9:   Review -- Pitfalls to Effective Implementation

STEP 1: Top Management Commitment

Without Chief Executive Officer's (CEO) commitment, no quality initiative can succeed. Where does this type of top management commitment come from? Many ISO 9000 registered companies find that the commitment comes from some, if not all of the following points.
- Direct marketplace pressure: requirements of crucial customers or parent conglomerates.
- Indirect marketplace pressure: increased quality levels and visibility among competitors.
- Growth ambitions: desire to exploit EC market opportunities.

- Personal belief in the value of quality as a goal and quality systems as a means of reaching that goal.

STEP 2: Establish Implementation Teams

ISO 9000 is implemented by people. The first phase of implementation calls for the commitment of top management - the CEO and perhaps a handful of other key people. The next step is to create a personnel structure to plan and oversee implementation.

The first component of this personnel structure is the Management Representative (**MR**). In the context of the standard, the MR is the person within the organisation who acts as interface between organisation management and the ISO 9000 registrar. His role is, in fact, much broader than that. The MR should also act as the organisation's **"quality system champion,"** and must be a person with:
1. total backing from the CEO,
2. genuine and passionate commitment to quality in general and the ISO 9000 quality system in particular,
3. the dignity - resulting from rank, seniority, or both - to influence managers and others of all levels and functions,
4. detailed knowledge of quality methods in general and ISO 9000 in particular.

Next, a top-level implementation team, known as the **Quality Steering Committee (QSC)**, is created. The team is usually chaired by the MR and consists of:
- The CEO
- Top managers
- Key functional managers
- Trade union representative, if any

The QSC is a group for formulating policies. It sets objectives for quality implementation, approves plans, evaluates reports, and prescribes changes as needed. The committee also makes critical decisions about the quality system documentation. Its members should decide on who is:
- responsible for writing, editing, and approving the organisation's quality manual, and
- responsible for second-tier and (if utilised) third-tier documentation.

Next comes a network of **Quality Action Teams (QATs)**. In small to medium-sized organisations, these teams are organised by department or function.

Each is headed by the functional manager or department head, who sits on the QSC. Larger organisations have more elaborate action networks, but the objective is the same: to have a team of line people representing each critical organisational function and process element. Furthermore, there must be a clear and visible reporting and communication network all the way up to the MR and the QSC.

STEP 3: Assess Current Quality System Status

ISO 9000 does not require duplication of effort, redundant systems, or make-work. The goal of ISO 9000 is to create a quality system that conforms to the standard. This does not preclude incorporating, adapting, and adding onto quality programmes already in place.

So the next step in the implementation process is to compare the **organisation's existing quality programmes -- and quality system,** if there is one -- with the requirements of the standard. Programme assessment can be done internally, if the knowledge level is there. Or a formal pre-assessment can be obtained from any one of a large number of ISO 9000 consulting, implementing, and registration firms.

STEP 4: Create a Documented Implementation Plan

Once the organisation has obtained a clear picture of how its quality system compares with the ISO 9000 standard, all nonconformances must be addressed with a documented implementation plan.

This plan may be created by an ad hoc committee under the authority of the QSC. Usually, the plan calls for setting up procedures to make the organisation's quality system fully in compliance with the standard. Procedures which affect high-level policy elements of the quality system may be handled by the council itself, or by designated members. Others may be handed down to various QATs for development.

The implementation plan should be **thorough and specific**, detailing:
1. Procedures to be developed
2. Objective of the system
3. Pertinent ISO 9000 section
4. Person or team responsible
5. Approval required
6. Training required
7. Resources required
8. Estimated completion date

These elements should be organised into a detailed **GANTT chart,** to be reviewed and approved by the QSC. Once approved, the plan and its GANTT chart should be controlled by the MR. The chart should be reviewed and updated at each QSC meeting as the implementation process proceeds.

As mentioned earlier, areas requiring implementation can vary widely from organisation to organisation, and depend upon the type and level of quality system elements already in place. Some organisations, with long-standing total quality management **(TQM)** or other quality systems, may need only rudimentary implementation measures in order to deal with nonconformances. Others may require the development of a quality systems from scratch -- a process which can take months, if not years.

Most organisations with some level of quality programme already in place find that their biggest areas of nonconformances are in:
- Design (for ISO 9001)
- Purchasing
- Inspection and testing
- Process control

STEP 5: Provide Training

The ISO 9000 implementation plan will make provision for training in various functional areas of the quality system. Certain training needs will depend on the nonconformances addressed. The QATs should take responsibility for providing specific training in their respective functional areas.

Since the ISO 9000 quality system affects all areas and all personnel in the organisation, it is wise to provide basic orientation in the quality system standard to all employees. This can be a **one-day programme which informs personnel about quality system in general and the ISO 9000 quality system in particular.**

The training programme should emphasise the benefits that the organisation expects to realise through its ISO 9000 quality system. The programme should also stress the higher levels of participation and self-direction that the quality system renders to employees. Such a focus will go far to enlist employee support and commitment.

STEP 6: Create Documentation

As noted earlier, documentation is the most common area of non-conformance among organisations wishing to implement ISO 9000 quality systems. As one company pointed out: "When we started our implementation, we found that documentation was inadequate. Even absent, in some areas. Take calibration. Obviously it's necessary, and obviously we do it, but it wasn't being documented. Another area was inspection and testing. We inspect and test practically every item that leaves here, but our documentation was inadequate."

There is no way around it: documentation is mandatory. It is essential to the ISO 9000 registration process because it provides objective evidence of the status of the quality system. The two basic rules of ISO documentation are:

- **Document what you do.**
- **Do what you document.**

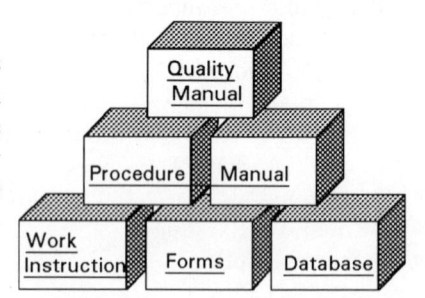

Many organisations find that their existing documentation is adequate in most respects. To bring it into full ISO conformance, they implement control procedures to ensure that documentation is available as needed and is reviewed, updated, stored, and disposed of in a planned, orderly manner.

In this context, documentation refers mainly to procedures. Under the ISO 9000 quality system, **all work which affects quality must be planned, controlled, and documented.** In other words, people whose work affects quality should know how to do that work in a way that maximises quality. Therefore, detailed written procedures and work instructions must be created where, as the standard states, the absence of same would adversely affect quality.

For many organisations, the process can become nightmarish. Don Van Hook observes: "One of the hardest things to do is to get shop people to become what they think of as bookkeepers. They don't like paper-work. To many of them it's a dirty word." The process of creating and using documentation is central to the effectiveness of quality system implementation. It is also educational.

Benefits of Documentation Include:

1. Quality system documentation forces the organisation, at all levels, to think through exactly what is being done and how it is being done. Most firms find this to be a positive experience:
    - It disseminates knowledge of the process.
    - It pools process knowledge and expertise.
    - It creates positive interfaces among individuals and process elements.

Chapter 7: ISO 9000 & Quality Audit 193

2. Through these interfaces, people develop positive, proactive approaches to teamwork and quality:
- They learn how to work together better.
- They learn what to expect of one another.
- They establish communication channels that result in positive improvement.

3. This is far more powerful, and far more important, than the mere process of writing procedures and work instructions. But writing them down is a vital part of the documentation process, too. Written documentation is:
- Evidence that thought has been given to the procedures.
- An irreplaceable reference resource for outside assessors.
- An invaluable training and improvement tool.

Keep in mind that quality system documentation does not need to be exhaustive, or redundant. It is certainly not intended to be an end in itself. Again, ISO 9000 requires only that the organisation have documentation -- procedures, work instructions, and the like -- the absence of which would adversely affect quality. **The organisation must maintain the minimal amount of documentation necessary to demonstrate that its quality systems exist and are being operated.**

STEP 7: Document Control

Once the necessary quality system documentation has been generated, a documented system must be created to control it. As noted in the Technical Requirements and Guidelines sections, control is simply a means of managing the creation, approval, distribution, revision, storage, and disposal of the various types of documentation. Document control systems should be as simple and as easy to operate as possible -- sufficient to meet ISO requirements and that is all.

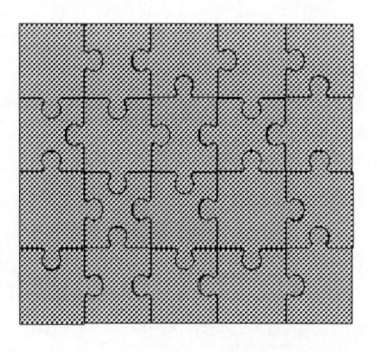

The principle of ISO 9000 document control is that employees should have access to the documentation and records needed to fulfil their responsibilities. Ironically, direct access can often result in certain employees having less record-keeping and documentation to deal with -- and can be a cause of resistance. The organisation's quality manual is a primary example. "We got minor resistance from some major players who were used to having the quality manual, but who didn't really need to have their own copy of it," says Jim Ecklein of Augustine Medical [Johnson, 1993]. "We solved that by having a master quality manual, with references to sub-manuals for each organisation area. That way, people had what they needed, but we weren't passing quality manuals out to people who didn't really need it and wouldn't use it."

STEP 8: Monitor Progress

When the procedures have been completed and the quality system fleshed out, it is time to put the quality system into effect. In this extremely important phase, management must pay close attention to results to make sure that the elements of the quality system are logical and effective.

Effective monitoring is what makes or breaks ISO 9000 implementation. It is also the ultimate measure of how well -- or poorly -- organisation management lives up to its responsibilities, as described in the Management Responsibility section of the standard. In particular, management at all levels should watch out for gaps and assumptions in procedures and steps which are difficult, ineffective, or impractical.

Many such problems can be dealt with by the QATs. Resulting changes should, of course, be documented and approved in accordance with procedures provided for in the quality system.

Management, up to the level of the Quality Action Council, should simultaneously carry out its review function as prescribed by the standard and its own documented procedures. These activities **include:**
- Internal quality audits
- Formal corrective actions
- Management reviews

It is especially helpful to begin the internal audit programme immediately as part of implementation, thereby solving many implementation problems at the "local" level. "We have internal audits every four weeks," says Anthony Coggeshall of Adhesives Research [Johnson, 1993]. "Each team consists of members of two unrelated departments -- not the quality department, and not the department being audited. These audits are really forums for departments to talk to one another and solve their own problems."

STEP 9: Review - Pitfalls to Effective Implementation

Here is a brief checklist of the most significant barriers to effective ISO 9000 quality system implementation.
1. ***Lack of CEO commitment.*** As Lorcan Mooney says, "If senior management consists of four or five people, and two of them are not committed, over time

they can be won over. But if the CEO is not committed, then in no way are you going to win in the long run."
2. *Failure to involve everyone in the process.* Ownership and empowerment are the keys to effective implementation. To help employees feel like owners of their activity, make them responsible for developing and documenting their procedures.
3. *Failure to monitor progress and enforce deadlines.* People in organisations have their routine work to do. If progress is not monitored, ISO 9000 can never be implemented effectively because programmes just drag on.

All three pitfalls are directly traceable to management -- or lack of it.

Tips on ISO 9000 Implementation

"We must focus on quality improvements and let quality system emerge based on improvement experiences."

-- Hans J. Bajaria, President, Multiface Inc.

Many companies find difficulty in the interpretation of the various sections of the standard. Chemoxy International Plc. [West, 1988] reported that they faced major disagreement in interpreting the independence of the management representative, the requirement of calibration and the extent of documentation needed. Fear of change at the start of the project was among the difficulties faced by the company during the implementation.

When effectively implemented, ISO 9000 quality systems provide positive benefits almost at once. "Of course we started with nothing." says Sandy Weller of Woodbridge Foam [Johnson, 1993]. "But almost from the start we noticed a major reduction in errors. Just the process of putting in the system, documenting what we were doing, and keeping track of what was going on improved quality -- because it made everyone aware of his or her impact on quality."

Benefits of Implementing ISO 9000

Bulled [1987] finds out that the potential benefit or advantage of implementing a quality system based on ISO 9000 fall into two categories: 1) the advantage of having the system, and 2) additional advantage accruing from the result of having a quality system that has been independently assessed.

The British Standards Institution in an effort to market ISO 9000 have published some of the claimed benefits [BSI, 1987]. They are:
1. BSI certification is a first class marketing tool. The certification marks and symbols can be used on publicity, packaging and company literature.
2. Major buyers like the Ministry of Defence and British Coal already accept BSI certification and registration as proof of quality and technical expertise.
3. Customers are much less likely to act for their own special assessments thus saving everyone's time and money.
4. Where there is a need for it, a company will improve its quality performance and as quality rises so will company morale.
5. The cost of lost orders, reworking, extra handling, production wastage, senior executive time will all come down once you are operating to ISO 9000.
6. Better quality performance will improve customer satisfaction and lead to increased sales, competitiveness and profitability.
7. The confidence that comes from knowing that your quality system is under independent surveillance.
8. The company's name will appear in the BSI Buyer Guide -- an essential reference book for buyers at home and abroad; and also in the Department of Trade and Industry's National Register of Quality Assessed Companies.
9. As more British Standard becomes harmonised with international ones, BSI certification will be of increasing help to the public in export markets.

Some case examples of the benefits of going for ISO 9000 are shown in Mini-Cases 7.2--7.6.

■■■

Mini-Case 7.2:

Benefits of ISO 9000 Implementation   [Straw 1988]

The Metal Finishing Industry has claimed tremendous benefits of implementing the standard. Some reported benefits are:
1. Better work flow through the factory, improved efficiency which leads to better customer service.
2. Consistent standard of training of new operators.
3. Consistent reduction of reworking resulted in savings of over £10,000.

Chapter 7: ISO 9000 & Quality Audit

4. The work-force is much more conscious of their contribution to the quality of work and became more involved in shop floor quality improvements.
5. Better document control which leads to improved internal and external communication.
6. Fall of customer rejects to only 40% of the previous level which results in quantifiable savings, repeated businesses, and a steady increase of new contracts.
7. In-house rejection rate has been halved, savings in the consumption of chemicals such as electrolyte and electrode are well over £40,000 annually.
8. Higher quality work is noticeable as the customer return falls from 3% of sales revenue to less than 0.5%. This results in savings of more than £30,000.

■■

■■

Mini-Case 7.3:

MBI's ISO 9002 Employee Attitude Survey

MBI has 20 employees and is mainly involved in the sale of industrial machinery and accessories to major manufacturing companies throughout Britain. It is part of a Group company with five other subsidiaries. From humble beginnings in 1970, MBI now supplies industrial tools to large companies such as British Telecom and British Steel. MBI is now a major player in their industry. In early 1993, BT de-listed MBI even though deliveries and service were agreed to be almost perfect. The reason was that BT's policy is not to do business with companies who are not ISO 9000 registered. Consequently, BT purchased from a third party, whom MBI supplied, that did have ISO 9000. As a result BT had to pay extra for the same product. In late 1993, MBI started to implement ISO 9002 into their operations and was accredited by the British Standards Institute by end-1994.

An interview was conducted with the Company Director, Senior Manager, Service Engineer, Sales Office Administrator, Warehouse Manager, Purchasing Manager and the Quality Representative to find out the attitudes of staff towards ISO 9002 implementation. The objective was to review the experience of MBI and consider it as a ISO 9000 implementation model for the rest of the Group. The findings/recommendations are summarised as follows:

1. *Take time over implementation; speed can cause employees many headaches and is likely to reduce their job satisfaction which in turn will reduce productivity.*
2. *Encourage staff participation; in general, if people are involved, their level of commitment is increased, as they feel in some way responsible.*

3. *Brief staff on company and individual aims and goals; if the idea of quality is sold successfully to the employees they are more likely to respond positively.*
4. *Give all staff the opportunity to give feedback at regular intervals.*
5. *Employees are more likely to be conducive to a system if they can see personal and company benefits; some form of profit sharing scheme has been identified by MBI as a way to encourage staff.*
6. *Ensure the employees feel that they are having the standard tailored to their individual needs.*
7. *Ensure management are fully aware of the standard; only if senior management are fully knowledgeable can they pass their knowledge downwards to employees.*
8. *Ensure the employees do not feel that additional work is being created; if necessary additional employees should be commissioned to spread the work load.*
9. *The quality representative must not be seen as the only 'expert' on ISO 9000; every employee should know the importance of their input.*
10. *Problems within the standard need to be worked out in an open and informal way, the employees must never feel that they are unable to complain -- there are always ways to change, that can be built into the standard.*

■■

■■

Mini-Case 7.4:

Bowater Containers Southern Goes for ISO 9000 [Dawson, 1988]

The benefits of the standard are not only received by the companies that implement the standard but also by their customers. Reed Corrugated Cases sees advantages both in cost and image in the market place. *Advantage to their customers was in the reduction of product price because the company was able to reduce cost and rejects.* Bowater Containers Southern claimed that benefits to customers are fourfold:
1. Regular checks of incoming goods can be replaced with a random audit.
2. Large amounts of stocks no longer need to be held as a safeguard against the quality of the supplier's next delivery.
3. Tight packaging specifications are maintained for clients that have their own automated plant.
4. Improved communications with all its suppliers.

■■

Mini-Case 7.5:

DTI Import Licensing Branch Goes for ISO 9000   [DTI, 1994, Mar.]

On 1 May 1991, the Import Licensing Branch (ILB) of the DTI was the first DTI division to be awarded a Certificate of Registration for ISO 9002. The ILB is one of six Government departments to issue import licenses.

Martin Maclean, Director of ILB, explains: 'With such a large scale of application processing, with mandatory targets and with interface with our customers being vital to our success, we were faced with a difficult task. The need for a quality standard was an important part of our response. In consultation with our staff we decided to go for ISO 9002 certification. The team spirit was high and the commitment of the staff so great that we achieved the standard in only 6 months.'

The process involved updating procedures, establishing and documenting an effective quality system and introducing a formal monitoring programme. The benefits for ILB have been a more disciplined approach, a greater concern for quality of service and a sound base for future improvements in efficiency.

Mini-Case 7.6:

Praxis Software Engineering Company Goes for ISO 9000   [DTI, 1994, Mar.]

Praxis is a software engineering company whose Bath office, in 1986, was the first independent software house to achieve certification to ISO 9001, and whose operating units in Warwick and Staines also achieved registration in 1990.

Praxis' approach is based on TQM aimed at securing client satisfaction and continuous improvement. Every year the Board identifies key aims and measurable objectives for the company. This approach is carried throughout the organisation with detailed objectives for each member of staff, who seek ways to improve their performance. If necessary, Quality Improvement Teams are established to tackle particular problems and to bring about improvements in process. When the process has been improved and its objectives are being met, the team updates the relevant written procedures before being disbanded, or re-directed to work on an even stiffer objective.

Chapter Summary:

- ISO 9000 series sets out the methods that can be implemented in an organisation to assure the customers' requirements are fully met.
- The content of ISO 9001 -- 20 clauses and sub-clauses which stipulates the conducts for a good quality management system. They can be grouped under the business functions of an organisation.
- To facilitate an quality audit, whether 1st, 2nd or 3rd party, there need to be a set of guidelines and quality auditors need to be trained accordingly.
- ISO 9000 is needed to demonstrate that a company is in control of its business, and has proved it to a certification body.
- Implementation of ISO 9000 is as simple as knowing the requirements, Understanding the organisation and its processes, committing the resource, planning well and seeing it through.
- The benefits of implementing ISO 9000 are plentiful, including the reduction of product price because the company was able to reduce cost and rejects.

Chapter 7: ISO 9000 & Quality Audit

Self-Assessment Exercise 7.1 : Interpretation of the ISO 9001 Clauses 4.1--4.20

1. When is sampling unacceptable?

2. Under what conditions should process/machine capability be known?

3. What section of ISO 9001 deals with the issue of materials and parts from stores?

4. When is a certificate of compliance not acceptable according to ISO 9001?

5. What section of ISO 9001 -- deals with:
   (a) Operator skill?
   (b) Operator control?

6. What section of ISO 9001 deals with the correct format for Test Certificates?

7. What does ISO 9001 have to say about productivity bonus systems?

8. What education should designers have?

9. What information is required for Design Reviews?

10. What procedures should be formal in a system to ISO 9001?

11. What types of companies do not have to have Inspection or Quality Control staff?

12. When are hand-written changes to documents allowed under ISO 9001?

13. How does ISO 9001 deal with quality costs?

14. Which part of ISO 9001 prescribes:
    (a) audits of suppliers?
    (b) a list of acceptable/unacceptable suppliers?

15. When does production equipment require calibration?

16. Which part of ISO 9001 deals with the actions to be taken in a company when a customer changes his order?

Self-Assessment Exercise 7.2 : ISO 9001 Quality Auditing

The auditors have discovered the following noncompliances. Write alongside "ISO 9001 Clause/Section" which Clause of the Standard the observed situation fails to comply with. If you think the auditors should not declare a discrepancy, write "N.A."

1. The copy of the Process Sheet X92/2/9 covering Hi Creme Lo Fat Rice Puddings lodged with the Shift Process Foreman was endorsed 'Not under amendment control'.
   ISO 9001 Clause/Section:
   Discrepancy (if any) :

2. In the Spice Room stocks of additives were found to have been contaminated by fumes from extraction ducts which were not airtight.
   ISO 9001 Clause/Section:
   Discrepancy (if any) :

3. A written instruction requires the involvement of Quality Control when labels are restored to goods which have lost their labels. The personnel who said that they restored labels were unaware of the authorised procedure and the procedure they described did not comply with it.
   ISO 9001 Clause/Section:
   Discrepancy (if any) :

4. No internal quality audit had been carried out on Personnel, Accounts, Sales, Administration, Data Processing or Quality. The Quality Manual (clause 13.6) states that audits will be carried out on all departments on a six monthly basis as a minimum.
   ISO 9001 Clause/Section:
   Discrepancy (if any) :

5. The Heading Master for component nos. AOP10296/B, CDF 14912/E and DSP18777/A did not communicate all the special manufacturing instructions for the relevant order 9OA694 due to lack of space on the card. The DP clerk had been given no guidance as to how to deal with such a contingency.
   ISO 9001 Clause/Section:
   Discrepancy (if any) :

Chapter 7: ISO 9000 & Quality Audit   203

6. A material critical to product quality is not covered by a purchase specification and there is no procedure for quality verification of incoming supplies.
   ISO 9001 Clause/Section: _____
   Discrepancy (if any) :

   _____

7. The Slave template used for checking boards on the production lines was badly maintained. Guide pins were worn causing misalignment and the reflection surface for underside inspection was very dirty.
   ISO 9001 Clause/Section: _____
   Discrepancy (if any) :

   _____

8. Although alterations to customers' orders are recorded on receipt in Sales, there is no method to ensure the changes are implemented throughout the system.
   ISO 9001 Clause/Section: _____
   Discrepancy (if any) :

   _____

9. It is a requirement that test bottles used for the Massey Checkweigher are approved and issued by the Laboratory. On the Hydraulic lines there were test bottles being used which had not been so approved.
   ISO 9001 Clause/Section: _____
   Discrepancy (if any) :

   _____

10. On the can filling line, the requirement of 50 cans per hour to be inspected was not being met. Between 5am and 12am an average of 10 cans per hour was inspected.
    ISO 9001 Clause/Section: _____
    Discrepancy (if any) :

    _____

11. Wax was being held longer than the 2 minutes allowed at pouring temperature and was not sieved before being applied on the line.
    ISO 9001 Clause/Section: _____
    Discrepancy (if any) :

    _____

12. Final Test records of seat leakage on control valves showed occasional hand-written changes. These lacked an authorising signature or date.
    ISO 9001 Clause/Section: _____
    Discrepancy (if any) :

    _____

Self-Assessment Exercise 7.3: ISO 9001 Mock Audit

1. The latter stages of the audit covered mainly production areas assembly, processing, test, calibration and despatch. Having explained their strategy and their likely whereabouts for the day to the Production Manager, the auditors set off accompanied by the Senior Foreman delegated to escort them. In the Assembly area the auditors walked towards the area for PCB assembly and component insertion, noting the standard of general housekeeping. Noticing a small area set aside from the main lines, the Team Leader asked the Foreman what it was. They were told it was used for 'one off' jobs and for 'difficult' repairs. Seeing some soldering being done in the area, the Team Leader walked over and asked the Foreman what was being done, The Foreman explained that Service Engineers carried out minor repairs on equipment from clients' sites. Noticing one engineer's toolkit open on the bench as well as car keys, jacket and lunch, the Team Leader asked if tills work was subject to ESD control. The Foreman answered that it was, but equipment had to work in sometimes hostile environments. With relief the Foreman pointed out the conductive mat on the floor and the ground cord attached to it and a bench mat. The Team Leader asked whether the controls should include the use of a wrist strap attached to earth and the 'caution' marking of the area as stipulated in the company's SOP501. The Foreman went to speak to the engineer who shrugged his shoulders and went on with his work. The Foreman returned and said the board had come in from a client's site and needed a very minor repair. The Team Leader again asked whether anti-static precautions should apply. The Foreman stated that they should but Service didn't come under Production and the engineers were very experienced and wouldn't take any undue risks. Walking over to another Engineer working on a board, the Foreman indicated both wrist strap & marking in evidence

   Nonconformance(s)   ISO 9001 Clause/Section
   _____   _____
   _____   _____

2. Continuing along the lines, the Team Leader asked to speak to one of the inspectors who was auditing in this area. He asked the inspector exactly what he did and listened as the Inspector explained. The auditor asked him for his Audit Plan. The Inspector said there was no written Audit Plan. The Auditor then asked how the Inspector knew what aspects he was required to check and their frequency. It was explained that when the operator came to a 'HP' (Hold Point) on the Master Process Sheet, the operator could only add their approval of their own work if the Inspector approved a sample first. HP was added to the sheets by Production Engineering on Engineering instruction. The operator pressed her warning light and the Inspectors made their way around to the various lights.

Chapter 7: ISO 9000 & Quality Audit   205

Nonconformance(s)   ISO 9001 Clause/Section
_____   _____
_____   _____

3. The Team Leader asked the Inspector how nonconformance was reported. The Inspector showed the auditor a book of nonconformance forms. Each form bore a serial number and they were completed in quadruplicate. In response to the auditor's question the Inspector explained that it was quite a recent system but the white copy was given to the chargehand, the second copy (pink) he attached to the work, the third copy (yellow), was meant to go to the Quality Office and the bottom copy (green) he kept for his own record. The Team Leader saw that the book started at no 261 and eight forms had been used. The yellow copies were still in the book. Form 261 was dated 16 days before and form 268 was dated the day before. He asked the Inspector what procedure was followed to get his copies to the Quality Office. The Inspector said there was no laid down procedure -- he would give one of the Quality Engineers the copies when he met him in the plant. The Inspector said he didn't know why the Quality Office was supposed to get copies of the reports.

Nonconformance(s)   ISO 9001 Clause/Section
_____   _____
_____   _____

4. The auditors asked the Inspector what indicated the 'inspection status' of the boards at the various stages. The Inspector pointed on the boards to the stamps annotated by the operators and also to his 'QA' stamp on the Process Sheet over the 'HP'. At the end of the lines the boards went through a series of tests for shorts, dry joints, continuity etc. and at this point they were given a yellow label. The Inspector pointed them out. The second man noted that one he could see was not signed and dated in the space provided. He referred to the Quality Manual (clause 12.6) to refresh his memory on test labels. He noted that all such labels were supposed to be signed and dated.

Nonconformance(s)   ISO 9001 Clause/Section
_____   _____
_____   _____

5. The audit party made its way towards the Test Shop and were introduced by the Assembly, Foreman to the Senior Test Foreman. The Team Leader explained to the foreman what they wished to discuss and see. Whilst the Team Leader, was talking to the Foreman, the second man was looking around and noted testing temperatures in a Test Schedule TS9669/B. The temperature was 80 deg. C for 10 hours. Realising this must be a special requirement he passed a note to the Team leader.

The Team leader acknowledged it and asked the Foreman if he could see the records of the temperatures and soak times for the extended temperature treatments. He was shown an A4 hardback duplicate book in which the contract number, unit number (Product identification), time, temperature, date were added. He noted that the temperature was always the same though the times varied with the equipment and date. A constant temperature of 48 deg. C was recorded. He asked the Foreman for the records pertaining to the testing of TS9669/b and was shown the entry in the book. It was the day before. The temperature recorded was 4,8 deg C and the time was 15 hours. The foreman said the treatments were basically equivalent; he said there was only one temperature booth and it was not practicable to alter the temperature setting for different products.

Nonconformance(s)   ISO 9001 Clause/Section
_____   _____
_____   _____

6. The Team Leader asked how the temperature controllers in the booth were checked. The Test Foreman asked an apprentice working on a module nearby to go and fetch the thermometer. He returned with a thermometer with a scale reading of 0-100 deg C. The auditor asked the apprentice if he knew about the use of the instrument. The apprentice said no. The Foreman said he personally checked the booth temperature against its setting every Monday morning. He kept a record of these checks in a notebook suspended on a string from a nail in the wall. The auditor looked at the book and saw that the foreman entered into it his initials and the date when he checked the booth. There were no omissions from the record.

Nonconformance(s)   ISO 9001 Clause/Section
_____   _____
_____   _____

7. Asked by the auditor if the thermometer was within the calibration control system, the Foreman said the instrument was new and he did not consider it would require to be calibrated until it was one year old.

Nonconformance(s)   ISO 9001 Clause/Section
_____   _____
_____   _____

8. Turning to the area of functional testing the Team Leader asked the Foreman what records were kept of tests. He was shown a number of filing cabinets. He was told there was no written procedure for the filing of records. Records were filed in a folder by Contract Number. No one was employed to do this - the Test Engineers and Technicians filed the records they generated. The

Chapter 7: ISO 9000 & Quality Audit

auditor asked for records of a product they knew had completed the tests and was shown a number of folders covering loop tests, sequential tests, system interaction, power failure etc. The auditors noted that the records in a number of folders were in date order but those in the third were not. The Foreman said someone had probably dropped this folder, but it would be sorted out before the records in it were checked as part of the final acceptance procedure.

Nonconformance(s)   ISO 9001 Clause/Section

9. The auditors asked the Foreman about the calibration of equipment. He led them to the Calibration Room and explained that all equipment was listed and records of calibration held in a file. The Team Leader asked if he might see a selection of certificates for a number of meters. These it was stated were returned to the maker annually for servicing and recalibration. On examining a number of certificates he found one which was dated 16 months before and no more recent certificates could be found in the file for this piece of equipment an ammeter. The auditor asked if any form of calibration programme was kept. As the Foreman replied that there was, he was called away to the telephone. Addressing a Calibration Technician the Team Leader asked him how they made sure they calibrated by the due date. The technician said they tried to go through the file periodically and arrange for servicing of equipment that was nearing to date. The Team Leader asked him where the ammeter was that had gone overdue. It was pointed out on a shelf behind two other units. The Technician said it had come back in the day before from a Service Engineer when he had come in to get a repair done. At that moment the Foreman returned, apologised to the Team Leader and stated that the period of calibration frequency wasn't critical and the ammeter had not been in use anyway.

Nonconformance(s)   ISO 9001 Clause/Section

10. The Team Leader asked how the equipment fixed out in the shops was calibrated. The Foreman explained the use of the various standards they held and the records produced. This was explained and verified to the auditors' satisfaction; the second man had taken identity numbers of various pieces of equipment and had checked these with the records. Asking about the calibration procedures, the Team Leader was handed a substantial file of documents which he examined. He found each procedure carried a reference number, revision number, date and authorising signature. He asked if the procedures were under Configuration Control as stated in the Manual and was told that this was the case. However their procedures were used by not only

the Calibration technicians but also Test and Design Engineers and by Plant Maintenance on occasions. The file copy was treated as the 'master' copy which was photocopied as required. The auditor asked if there was a register of the photocopies and was told 'no'. However, when a procedure changed, which was very infrequent, all the noted departments were advised by memo. All copies of the procedures were also date stamped and although it wasn't written down anywhere, everyone knew that any calibration procedure over 3 months old or so should be checked for currency.

Nonconformance(s)   ISO 9001 Clause/Section
_____   _____
_____   _____

11. Moving on from the Calibration Room the audit party walked through the Final Viewing Area. This area (under the responsibility of Testing) was stated to be the area which ensured that the full requirements had been catered for. Examinations were made of all visual aspects inside and out of all equipment, documentation was collated, any last minor 'snag' items were corrected, surfaces were cleaned if necessary and special protective devices were added where necessary.

The auditors asked how status was shown on finally accepted items and were pointed to a 'Final View OK' label in the company colours attached next to the serial number of the unit. Passing by a large tower unit with the door open, the second man commented on the fact that cable locators appeared very sparse and the door itself was likely to foul internal equipment when it was closed and opened. The Foreman looked at the unit and pointed out the 'Final View OK' label, restating that it must be all right. The auditor however asked if he could check that the appropriate specification be raised from Product Development to check the Layout.

Nonconformance(s)   ISO 9001 Clause/Section
_____   _____
_____   _____

12. While waiting for the specification the auditors asked if they could be taken to the Despatch Area, and return in an hour to satisfy the point. The Test Foreman took them to the Despatch Area and introduced them to the Chargehand. He asked that they wait a little if they didn't mind because he had been instructed by the Logistics Manager to tell him immediately the auditors appeared, so that he could deal with them personally. The Manager arrived promptly and ensured the auditors were given assistance and information as necessary. Procedures for packing and identification etc. were seen to be in order and the area was clean and tidy. On the way out of the Despatch Area, the Manager stopped to look at five units in packing cases (serial nos. CZ 116, CZ 119, CZ 120, CZ 141, CZ 142). He called over the

Chapter 7: ISO 9000 & Quality Audit

packer and asked him why the units were packed with corrugated cardboard instead of expanded polystyrene foam. The packer replied that the Chargehand had said he should use cardboard when they ran out of polystyrene foam two weeks ago. He pointed to the typed Engineering Procedure Instruction (ENG 194) on the wall above the bench. There was a hand-written amendment to the instruction about the packing which had been initialled 'PCH' and dated 4 months previously. In response to the auditors' query the Manager stated these were the initials of the Chargehand.

Nonconformance(s)   ISO 9001 Clause/Section
_____   _____
_____   _____

13. Returning to the test area, the Team Leader asked to see the specification which was now available. Support spacings were 50 cm max. and out of 12 spacings, 5 were 70-100 cm. The door have been side hinged instead of bottom hinged. The Foreman told them it will have been 'human error' and would be picked up by the Service and Customer Support staff on site. The auditor asked to see evidence that the Test personnel had examined the unit. He was told the Technician who had done the work was at a client's site all that week. The second man reminded the Team Leader that they were going to be late for the next stage of the audit. Thanking the Foreman the auditors left the area.

Nonconformance(s)   ISO 9001 Clause/Section
_____   _____
_____   _____

Hands-on Exercise:

1. Follow the nine-step procedure, formulate a plan to implement ISO 9000 in you own organisation.

2. How can you build ISO 9000 quality management system into your TQM implementation programme?

Annex 7.1 : List of the ISO 9000 Family

[1] ISO 8402:1994, *Quality management and quality assurance -- Vocabulary.*
[2] ISO 9000-1:1994, *Quality management and quality assurance standards -- Part 1: Guidelines for selection and use.*
[3] ISO 9000-2:1993, *Quality management and quality assurance standards -- Part 2: Generic guidelines for the application of ISO 9001, ISO 9002 and ISO 9003.*
[4] ISO 9000-3:1991, *Quality management and quality assurance standards -- Part 3: Guidelines for the application of ISO 9001 to the development, supply and maintenance of software.*
[5] ISO 9000-4:1993, *Quality management and quality assurance standards -- Part 4: Guide to dependability programme management.*
[6] ISO 9001:1994, *Quality systems -- Model for quality assurance in design, development, production, installation and servicing.*
[7] ISO 9002:1994, *Quality systems -- Model for quality assurance in production, installation and servicing.*
[8] ISO 9003:1994, *Quality systems -- Model for quality assurance in final inspection and test.*
[9] ISO 9004-1:1994, *Quality management and quality systems elements -- Part 1: Guidelines for services.*
[10] ISO 9004-1:1991, *Quality management and quality systems elements -- Part 2: Guidelines for processed materials.*
[11] ISO 9004-1:1993, *Quality management and quality systems elements -- Part 3: Guidelines.*
[12] ISO 9004-1:1994, *Quality management and quality systems elements -- Part 4: Guidelines for quality improvement.*
[13] ISO 10011-1:1990, *Guidelines for auditing quality systems -- Part 1: Auditing.*
[14] ISO 10011-2:1991, *Guidelines for auditing quality systems -- Part 2: Qualification criteria for quality systems auditors.*
[15] ISO 10011-3:1991, *Guidelines for auditing quality systems -- Part 3: Management of audit programmes.*
[16] ISO 10012-1:1992, *Quality assurance requirements for measuring equipment -- Part 1: Metrological confirmation system for measuring equipment.*
[17] ISO 10013:1995, *Guidelines for developing quality manuals.*

Annex 7.2 : The 20 Clauses of ISO 9001:1994

4.1 Management responsibility
    4.1.1 Quality policy
    4.1.2 Organisation
        4.1.2.1 Responsibility and authority
        4.1.2.2 Resources
        4.1.2.3 Management representative
    4.1.3 Management review
4.2 Quality system
    4.2.1 General
    4.2.2 Quality system procedures
    4.2.3 Quality planning
4.3 Contract review
    4.3.1 General
    4.3.2 Review
    4.3.3 Amendment to a contract
    4.3.4 Records
4.4 Design control
    4.4.1 General
    4.4.2 Design and development planning
    4.4.3 Organisational and technical interfaces
    4.4.4 Design input
    4.4.5 Design output
    4.4.6 Design review
    4.4.7 Design verification
    4.4.8 Design validation
    4.4.9 Design changes
4.5 Document and data control
    4.5.1 General
    4.5.2 Document and data approval and issue
    4.5.3 Document and data changes
4.6 Purchasing
    4.6.1 General
    4.6.2 Evaluation of subcontractors
    4.6.3 Purchasing data
    4.6.4 Verification of purchased product
        4.6.4.1 Supplier verification at subcontractor's premises
        4.6.4.2 Customer verification of subcontracted product
4.7 Control of customer-supplied product
4.8 Product identification and traceability
4.9 Process control
4.10 Inspection and testing
    4.10.1 General
    4.10.2 Receiving inspection and testing

　　　　4.10.3　In-process inspection and testing
　　　　4.10.4　Final inspection and testing
　　　　4.10.5　Inspection and test records
4.11　Control of inspection, measuring and test equipment
　　　　4.11.1　General
　　　　4.11.2　Control procedure
4.12　Inspection and test status
4.13　Control of nonconforming product
　　　　4.13.1　General
　　　　4.13.2　Review and disposition of nonconforming product
4.14　Corrective action
　　　　4.14.1　General
　　　　4.14.2　Corrective action
　　　　4.14.3　Preventive action
4.15　Handling, storage, packaging and delivery
　　　　4.15.1　General
　　　　4.15.2　Handling
　　　　4.15.3　Storage
　　　　4.15.4　Packaging
　　　　4.15.5　Preservation
　　　　4.15.6　Delivery
4.16　Control of quality records
4.17　Internal quality audits
4.18　Training
4.19　Servicing
4.20　Statistical techniques
　　　　4.20.1　Identification of need
　　　　4.20.2　Procedures

*Chapter 8

Total Productive Maintenance

Topics:

► *What is TPM?*
► *The content of TPM*
► *Why is TPM useful?*
► *How to implement TPM*

```
   ┌─────┐
   │ 5-S │
   └──┬──┘
      ▼
   ┌─────┐
   │ BPR │
   └──┬──┘
   ......
      ▼
   ┌─────┐
   │QCCs │
   └──┬──┘
      ▼
   ┌─────┐
   │ ISO │
   └──┬──┘
      ▼
   ┌─────┐
   │ TPM │
   └──┬──┘
      ▼
   ┌─────┐
   │ TQM │
   └─────┘
```

Introduction: What is TPM?

Total Productive Maintenance, defined in Chapter 3, has its own history. It all started with Preventive Maintenance (PM) and referred to maintenance activities to prevent stoppages. It was expanded to include efforts to prevent quality degradation caused by equipment deterioration and malfunctioning. This additional responsibility was also within the scope of the maintenance department activities. Problems, obviously, are inevitable after the machinery or office equipment has been used for some time. It has been realised that the operators -- the people actually using the equipment who know the equipment best -- should be involved in the process of maintenance, repairs and changes.

PM has made it possible to improve the equipment to produce higher quality, and has thus become an important part of equipment control. It has also deterred the proliferation of undesirable features (e.g., features that endanger the worker or that pollute the environment). As a result, the original aspect of PM was upgraded into 'Total Productive Maintenance'. Furthermore, some successful organisations, such as Aishin Seiki, the Productive Maintenance Excellence Award winner for 1982 prefer the name 'Total Productive Management' indicating a step forward.

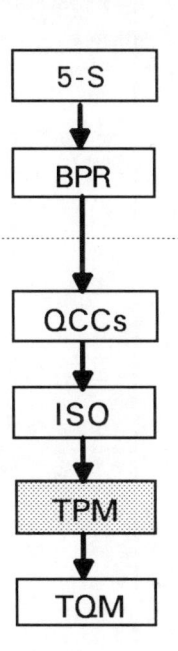

In the light of its contemporary context, productive maintenance cannot be an exclusive responsibility of the maintenance department. The broadened definition of PM emphasises on-site experience and calls for proactive solutions -- 'prevention is better than cure'. As a result, the maintenance department has to work with the production teams and to involve the planning department, human resources development, accounting, and more, if not all departments. TPM has evolved from being a specialist field for a small number of experts to being a generalised concern involving nearly everyone in the company, hence becoming *total* [Senju, 1992]. Figure 8.1 illustrates the theme of TPM, i.e., *prevention is better than cure*.

Figure 8.1 Prevention is Better than Cure (Rule TPM)
[Lip, 1989]

TPM involves everyone from the top executives to the shop floor workers to promote productive maintenance through morale-building management and small-group activities in an effort to maximise equipment efficiency.

Many people undermine the importance of small-group activities. Basically, TPM implies utilising plant capability to its fullest extent to:
- reduce equipment stoppages (both line stoppages and stoppages for reworking),
- quantitatively and qualitatively enhance equipment capability, and
- improve safety, health, and environmental factors in the expectation that such improvements will contribute to better quality and higher profits.
- utilise small-group activities and prevention.

These applications show that TPM process is a natural extension of the 5-S practices and QCC activities applied to equipment improvement. This also explains why the Japanese consider the 5-S and QCC the base for TPM implementation and an important stage in the development of TQM.

Chapter 8: Total Productive Maintenance (TPM)   215

The Content of TPM

Sponsored by the JIPE, the Productive Maintenance Excellence Award *recognises outstanding achievements in the area of TPM in Japan.* There are three levels of the Award: Level 1, Level 2, and Special Prize. The Level 1 Award is open to all companies regardless of size. The Level 2 Award is restricted to smaller manufacturing companies capitalised at 500 million Yen or less and having under 500 employees. The Special Prize is limited to entrants that won a Level 1 Award three or more years earlier.

The Productive Maintenance Excellence Award 10-point Checksheet developed by the JIPE gives a full insight into the context of TPM. The Checksheet encompasses the following ten points.

1. **Policy and objectives**
1.1 How is equipment management integrated into company policy?
1.2 Are equipment management policy and objectives set and prioritised correctly?
1.3 Are the management guidelines and evaluation criteria good?
1.4 Are long-term and annual plans integrated?
1.5 Are policy and objectives thoroughly understood and implemented?
1.6 Are accurate checks done to make sure objectives are being met?
1.7 Are the year's results considered in formulating goals, objectives, and plans for the next year?

2. **Organisation and operation**
2.1 Are the organisation and personnel assignments right for managing the equipment?
2.2 Is a good organisation in place for promoting TPM?
2.3 Is the TPM promotion organisation in close contact with production lines?
2.4 Are all the necessary departments participating fully in TPM?
2.5 Is effective co-ordination being done between the head office and the factories?
2.6 Are there any impediments to the exchange and use of information?
2.7 Are relations good with any subcontractors responsible for equipment, dies, tools, and maintenance work?

3. **Small-group activities and autonomous maintenance**
3.1 Is the organisation of the small groups right?
3.2 How are small-group activity objectives set?
3.3 Do the groups meet frequently and are their meetings lively?
3.4 Are there lots of good suggestions, and are they handled properly?
3.5 How is goal attainment ascertained?
3.6 Do operators take the initiative in maintaining their equipment?

4. **Training**
4.1 Do the different departments understand TPM?
4.2 Are the training programmes broad enough and their curricula right?
4.3 Do the training programmes follow the curricula?
4.4 Do people take part in outside training programmes?
4.5 How many people have technical or expert certifications?
4.6 How knowledgeable and skilled are people in maintenance techniques?
4.7 How is skill assessment done?
4.8 Is there any way to make sure the training is having an effect?

5. **Equipment maintenance**
5.1 Are the 5-S being implemented?
   - Is the machinery clean and free of dust, filings, oil, and other waste material?
   - Have policies been instituted to deal with dirt, places that cause soiling, and places that are difficult to clean, inspect, and lubricate?
   - Are lubrication labels, gauge limitations, bolt tighten-to-her marks, and other visible indicators used?
   - Are all of the tools, materials, gauges, and other things stored neatly and kept clean?
5.2 Are equipment diagnosis techniques used in these cases?
   - Cracking, corrosion, and loosening?
   - Abnormal vibration, heating, and noise?
   - Water, oil, gas, and air leakage?
5.3 Are power lines, water lines, hydraulic lines, and other lines neatly and properly handled?
5.4 Are oils properly selected and properly replaced or filtered at the appropriate intervals?

6. **Planning and management**
6.1 Are the appropriate efforts being made to improve maintenance techniques and efficiently?
6.2 Are equipment standards properly set and enforced in a planned manner?
6.3 Are monthly and annual maintenance plans drawn up and implemented properly?
6.4 Are the purchasing plans for spare parts and other maintenance equipment properly drawn up (e.g., how much of what to buy from where) and are such things cared for?
6.5 Are equipment blueprints well cared for?
6.6 Are the dies, tools, and gauges properly cared for?
6.7 Are good records being kept on equipment wear and equipment failures that mandate stoppages or other maintenance efforts?
6.8 Are maintenance records used to improve processes?
6.9 Are the right maintenance techniques properly applied?

Chapter 8: Total Productive Maintenance (TPM)

7. Equipment investment plans and maintenance planning
7.1 Is equipment investment co-ordinated with the development of new products and new processes?
7.2 Is equipment investment cost-effective?
7.3 How is the plant investment budget drawn up and controlled?
7.4 Are maintenance planning suggestions duly reflected in equipment investment standards?
7.5 Are reliability and maintenance duly considered in selecting, designing, and placing equipment?
7.6 Is equipment closely monitored in the start-up stages?
7.7 Is the company good at developing its own dies, tools, and equipment?
7.8 Are policies promptly instituted to keep major problems from recurring?
7.9 Are plant assets properly managed?

8. Production volume, scheduling, quality, and cost
8.1 Is equipment control closely integrated with production volume and scheduling?
8.2 Is equipment control closely integrated with quality control?
8.3 Are maintenance budgets drawn up and managed properly?
8.4 Is energy and other resource conservation practised?

9. Safety, sanitation, and environmental conservation
9.1 Are sound policies in place for safety, sanitation, and environmental conservation?
9.2 Are the right organisations in place for safety, sanitation, and environmental conservation?
9.3 Are safety, sanitation, and environmental conservation methods known and practised?
9.4 Is equipment investment integrated with safety, sanitation, and environmental conservation considerations?
9.5 Are safety and sanitation policies paying off?
9.6 Do environmental policies meet all of the legal requirements?

10. Results and assessments
10.1 Are results properly measured?
10.2 Are policies being implemented and objectives met?
10.3 Is maintenance paying off in terms of enhanced productivity and other management aims?
10.4 Does the company make an effort to publicise its activities and its successes?
10.5 Does the company know where the problems are?
10.6 Have plans been drawn up for future progress?

■■

Mini-Case 8.1:

1986 PM Excellence Award Winner: Toyota Auto Body Co. [Senju, 1992]

According to the President of Toyota Auto Body, 'Quality first' is a central tenet of his company's basic company-wide operating policy. Management has always put quality at the top of its priority list. These efforts resulted in Toyota Auto Body's receiving the Deming Prize in 1970, the Japan Quality Control Prize in 1980, and the PM Excellence Award in 1986. In order to achieve the PM Excellence Award, Toyota adopted a TPM programme to improve the management of equipment and an OPM (Office Productivity Maximisation) programme to streamline paperwork in the administrative and staff divisions.

The following points were emphasised in the TPM programme by improving process capability, eliminating all losses, and raising operating rates, to create a more productive assembly line that can turn out reliable products that will win customer trust:
1. *Simplify processes so that quality is not impaired by worker reassignments.*
2. *Build equipment so that lower production volumes do not result in higher per-vehicle energy consumption.*
3. *Build flexible equipment that can be adapted to model changes with only minimum investment.*
4. *Improve equipment reliability and life despite the greater automation and complexity.*
5. *Develop production technologies that solve these problems and reduce costs at the same time.*

A programme of OPM Streamlining staff operations was adopted that sought to improve the quality of work to make the entire system more efficient by returning to the basics, defining work goals clearly, and reorganising the office system in line with the concept of purposeful work. The OPM activities sought two goals:
1. *To improve the quality and efficiency of administrative and staff work.*
2. *To reassign staff to strengthen technology development and in anticipation of new work requirements.*

■■

Mini-Case 8.2:

1984 PM Excellence Award Winner: Fujikoshi [Senju, 1992]

While the trend seems to go from TQM to TPM, Fujikoshi has done the opposite. According to the president of Fujikoshi, Mr. K. Owada, the company's guiding principle is to ensure that they are producing quality products that meet the changing needs of society. As a general machine manufacturer, Fujikoshi has to be involved in the most basic fields of the machine industry, and it seeks to develop machine designs, processing technology, production systems, and new materials. Its main factory has about 5,000 employees. The company also has 35 affiliated sales and production companies with about 3,000 employees.

Particularly impressive was the company-wide human resources training initiative ranging from top management to the programme promotion staff and the front-line workers in the factories. It was this initiative that enabled Fujikoshi to implement such a sophisticated TPM programme so successfully. The programme brought out all of the workers' potential, and this should be invaluable for strengthening the company and promoting future developments.

Fujikoshi uses five basic policies incorporating both the company's operating policies and the problems discovered during the preliminary survey were announced for the TPM kick-off:

1. Raise overall equipment efficiency as much as possible through the full participation of all employees.
2. Enhance equipment reliability and maintainability, and seeking to build quality into the equipment and to improve productivity.
3. Develop the most economical and efficient equipment, and equipment management for the entire life of the equipment.
4. Train people to be competent with the equipment.
5. Create a lively work environment through autonomous maintenance.

While the trend is to go from TQM to TPM, Fujikoshi has done the opposite. As they developed quality education and training, the innovation suggestion system, and QCC, there was a tendency from many of the programme's vocal promoters to avoid the dirty work. this was disastrous for company morale. By implementing the TPM programme early on, Fujikoshi had all employees make actual improvements on the equipment and get their hands dirty. Their TPM programme resulted in cutting the number of stoppages to 1/150 of their former level, raising equipment efficiency by more than 30%, and reducing nonconformance to 1/3 of what it had been. Along with this, value-added productivity improved by 30%.

Why is TPM useful?

In modern day manufacturing and service industries, improved quality of products and services increasingly depend on the features and conditions of organisations' equipment and facilities. In the late 70's, there was heavy snow in Sapporo, the northern-most island of Japan. Because the workers could not get to work, Matsushita's vacuum cleaner factory stood still. Mr. Matsushita thought, 'Can we not rely on our workers for production?' A year later, the first unmanned-factory in the world was born. As the production relied 100% on equipment, TPM became mandatory.

Today, there are many similar examples such as Fujitsu-Fanuc, the world's most advanced unmanned-factory, which uses reliable computer controllers for manufacturing automation. Likewise super-computers run 24 hours a day all over the world to provide uninterrupted services to the banking, finance, air-flight, hotel, tourist, telecommunication and other service industries. However, this would not be possible without TPM.

TPM is a programme for fundamental improvement that involves the entire human resource. When implemented fully, TPM *dramatically improves productivity and quality and reduces costs.* As automation and labour-saving equipment take production tasks away from humans, the condition of production and office equipment increasingly affects output, quality, cost, delivery, health and safety, and employee morale. In a typical factory, however, many pieces of equipment are poorly maintained. Neglected equipment results in chronic losses and time wasted on finding and treating the causes.

Equipment Effectiveness is Everyone's Responsibility

Both operations and maintenance departments should accept responsibility of keeping equipment in good conditions. To eliminate the waste and losses hidden in a typical factory environment, we must acknowledge the central role of workers in managing the production process. No matter how thoroughly plants are automated or how many robots are installed, *people* are ultimately responsible for equipment operation and maintenance. Every aspect of a machine's performance, whether good or bad, can be traced back to a human act or omission. Therefore no matter how advanced the technology is, people play a key role in maintaining the optimum performance of the equipment.

When company employees accept this point of view, they will see the advantage of building quality into equipment and building an environment that prevents equipment and tools from generating production or quality problems. This company-wide team-based effort is the heart of TPM. It represents a dramatic change from the traditional "I make -- you fix" attitude that so often divides workers. Through TPM, everyone co-operates to maintain equipment the company depends on for survival and ultimately for profitability.

Chapter 8: Total Productive Maintenance (TPM)

Goals and Objectives of TPM

The goal of TPM is to increase the productivity of plant and equipment. Consequently, maximised output will be achieved through the effort of minimising input -- improving and maintaining equipment at optimal levels to reduce its life cycle cost. Cost-effectiveness is a result of an organisation's ability to eliminate the causes of the **'six big losses'** that reduce equipment effectiveness:
- Reduced yield (from start-up to stable production)
- Process defects
- Reduced speed
- Idling and minor stoppages
- Set-up and adjustment
- Equipment failure

Apart from these, TPM deals with other areas that needs improvement as given in Figure 8.2.

Work Time Breakdown

Before implementing TPM, it is always useful to collect some statistics for the utilisation of a piece of equipment first. This is necessary because a lot of facts about losses can be revealed through data analysis (e.g. Pareto Analysis in Chapter 6). A typical example of time breakdown for a piece of equipment is shown in Figure 8.2. The **Utilisation Rate** is the ratio of Operating Time (the first row in Figure 8.2) to the Total Available Time (the sum of all times in Figure 8.2).

Figure 8.2 implies that *the average equipment utilisation is usually very low.* Actions to reduce the 'Six Big Losses' and other types of losses are suggested in the right hand column of Figure 8.2. These losses deserve careful consideration as their elimination means productivity improvement, and quality improvement as well.

Time Element	Details of Activity	Possible Improvements
Operating time	• Number of non-defective products X standard cycle time	Design for manufacture
The Six Big Losses:		
Launch / Yield loss	• Time used for setting the condition, trial manufacturing, commissioning, etc., when starting up production or resetting after breakdown or shutdown	Preventive maintenance analysis
Defective / readjustment loss	• Time needed to manufacture defective goods in normal production • Time needed to stop the facility in order to select or readjust defective goods	Skill analysis
Speed reduction loss	• Time lost due to the difference between the standard processing speed and the actual processing speed	Identify quality conditions
Short-term shutdown loss	• Shutdown time due to frequent short-term shutdown of facility	Eliminate minor defects
Setting loss	• Time for replacing, adjusting, testing or processing dies, jigs and tools	Search optimal condition
Failure shutdown loss	• Time for repairing the cause of a shutdown due to an unexpected failure	Restore immediately
Further Losses due to production scheduling, planned maintenance, interruptions and rest-time:		
No load time	• Idle time due to the lack of work-in-progress • Time for awaiting the arrival of materials due to delayed delivery from subcontractors or other sources	Just-in-time production system
Planned stop time	• Time for planned preventive maintenance or corrective maintenance	Preventive engineering
Out-of-control time	• Shutdown time of facility due to a morning meeting, report meeting, participation in symposium, educational training, fire drill, health check, preventive injection, stock-take, shutdown of power, energy, shutdown, etc.	Better scheduling if possible
Rest time	• Rest time for operators influencing the operating time of the facility • Rest time determined by the production plan	Over-time work

Figure 8.2 Work Time Breakdown

How to Implement TPM

Implementation plans for TPM vary from company to company depending on the level of maintenance and particular plant requirements. TPM consists of **six major activities:**
1. *Elimination of six big losses* based on project teams organised by the production, maintenance, and plant engineering departments.
2. *Planned maintenance* carried out by the maintenance department
3. *Autonomous maintenance* carried out by the production department in seven steps.
    - **Step 1:** Initial cleaning
    - **Step 2:** Actions to address the causes and effects of dust and dirt
    - **Step 3:** Cleaning and lubrication standards
    - **Step 4:** General inspection training
    - **Step 5:** Autonomous inspection
    - **Step 6:** Workplace organisation standards
    - **Step 7:** Full implementation of autonomous maintenance
4. *Preventive engineering* carried out mainly by the plant engineering department
5. *Easy-to-manufacture product design* carried out mainly by the product design department
6. *Education and training* to support the above activities

TPM can be successful in achieving significant results only with universal co-operation among all constituents involved with the six activities listed above. Once a decision has been made to initiate TPM, company and factory leadership should promote all six of these activities despite excuses that may come from various quarters.

Through these activities, the company can gradually eliminate the losses shown in Figure 8.2, establish a more effective relationship between operators and machines, and maintain equipment in the best possible condition.

Prioritise Equipment TPM Rating

Equipment should be ranked according to the urgency of maintenance required. Figure 8.3 is a Worksheet to evaluate the TPM Rating of an equipment. It is designed to be useful for both factory and office equipment. Some of the ratings are fairly subjective, i.e. comparative. There is a need for an established set of measures through data gathering and experience. Nevertheless, if the worksheet is applied consistently on a similar range of equipment, the rating would be more meaningful. Once the ratings of a range of equipment are evaluated, then they can be prioritised for TPM implementation.

AREA	ITEM	Evaluation 1=====>5 low.....high
Production/ Operations	1. How often is the equipment used? 2. Is there backup equipment? (1=always there, 5=no) 3. What is the utilisation rate? 4. To what extent will a failure affect other processes.	
Quality	5. Value of monthly scrap and/or losses 6. The effect of the process on the quality of finished product/service	
Maintenance	7. Frequency of failures in terms of cost of repair. 8. Mean time to repair	
Safety	9. To what extent does a failure affect the work environment?	
	Total Points	

Figure 8.3 Equipment TPM Rating Table
(The higher the total points, the higher the priority for TPM.)

Chapter Summary

► *TPM is a system of maintenance covering the entire life of the equipment in every division.*
► *The JIPE TPM annual award recognises outstanding achievement.*
► *TPM is useful because it dramatically improves productivity and quality and reduces costs.*
► *From the work time breakdown it is not surprising that the average equipment utilisation is usually very low.*
► *To implement TPM follow the six major activities and the seven steps in Autonomous Maintenance, with the objectives to reduce all possible losses.*

Hands-on Exercise:

1. Select a piece of equipment which you are responsible for. Collect the data on its utilisation during a month. Tabulate the results in the form of work time breakdown. Do a Pareto Analysis to identify the priorities of the losses.

2. Apply the 'Possible Improvements' column of Figure 8.2 to minimise the losses you recorded in 1. above.

3. Follow the *six major activities and the seven steps in Autonomous Maintenance,* formulate a plan to implement TPM in your own organisation.

Chapter 9

TQM Kitemarks

Topics:

- What are TQM kitemarks?
- The content of ISO 9004-4
- Japan's Deming Prize
- USA's Malcolm Baldrige National Quality Award
- The European and UK Quality Awards
- Why are TQM kitemarks useful?
- How to achieve TQM kitemarks

```
5-S
 ↓
BPR
 ↓
QCCs
 ↓
ISO
 ↓
TPM
 ↓
TQM
```

Introduction: What are TQM Kitemarks?

Once you have gone through all the five stages of the TQMEX model, your organisation is likely to have built a very strong basis and a proactive environment for the final stage -- TQM. There are clear business objectives and effective processes installed, with empowered employees committed to quality; ISO 9000 quality management system is in place to demonstrate the disciplined approach to quality improvement and all equipment and facilities are in good condition and utilisation consistently. This is the right time for the management of an organisation to consider obtaining a nationally or internationally recognised quality standard registration.

TQM kitemarks are nationally or internationally recognised quality standards that provide discipline, external assessment, and a clear process for switching to TQM. They also have tremendous potential publicity value within the organisation and with the general public. Formal registration or award convey *important messages to actual and potential customers, that the institution takes quality seriously, and that its policies and practices conform to national and international standards of quality.* This can provide considerable external confidence as well as increase internal pride. The best known TQM kitemarks are the ISO 9004-4, Japan's Deming Prize, USA's Malcolm Baldrige National Quality Award, and the European and UK Quality Awards.

ISO 9004-4:1993

In 1992, the British Standards Institution developed the BS7850 guideline on TQM. *It is a guideline for quality improvement, with emphasis on management commitment and company-wide involvement.* BS7850 is not a Standard, and it is not envisaged to have a system of third-party assessment like ISO 9000. It is a guide to the type of approach and methodology which an organisation pursuing TQM might adopt.

In 1993, the International Organisation for Standardisation published a similar guide as part of the ISO 9000 series called ISO 9004-4 "Quality management and quality system elements -- Part 4: Guidelines for quality improvement". In 1994, the BSI revised its BS7850:1992 in accordance with the ISO 9004-4:1993.

ISO 9004-4 emphasises the participation and co-operation of all employees in the production of goods or services. The purpose is to ensure that customers are satisfied with the goods and services and that society as a whole benefits.

The 'Forward' to the Standard has an interesting section on the initial investment needed for TQM and it includes investment in time to train people, time to implement new concepts, time for people to recognise the benefits, and time for people to move forward into new and different company cultures.

The most important parts of the ISO 9004-4, Sections 4, 5, 6 and 7 and their sub-headings are shown in **Annex 9.1**. They focus on the fundamental concepts of management and methodology for quality improvement. A list of supporting tools and techniques which encompasses all the seven QC Tools discussed in Chapter 6 is also included.

Japan's Deming Prize

The *best-known prize, with the longest history*, is the Deming Prize awarded by the Japanese Union of Scientists and Engineers (JUSE) *to companies with outstanding TQM*. This prize is given for the overall performance of the company. There are, however, provisions for corporate divisions to compete for the prize. If a company has ten factories, then these ten factories, and of course the main office, are surveyed as one unit. Research centres and sales outlets are included, and random sampling is used if there are a large number of sales outlets. Five years after a company has received the Deming Prize, it is eligible to compete for the Japan Quality Control Prize.

The content of the Deming Prize is best understood by studying the 10-point Deming Prize Checksheet (1984 version) as follows:-

1. Policy
1.1 Management, quality, and quality control policy
1.2 Policy-making method
1.3 Policy correctness and consistency
1.4 Use of statistical methods
1.5 Policy transmission and dissemination

Chapter 9: TQM Kitemarks

1.6 Checking policy and attainment
1.7 Relation between long-term and short-term policies
2. **Organisation and Operation**
2.1 Clarification of authority
2.2 Appropriateness of delegation of power
2.3 Co-operation among divisions
2.4 Committee activities
2.5 Use of staff
2.6 Quality control audits
3. **Education and Training**
3.1 Plans and actual accomplishments
3.2 Quality-mindedness, control-mindedness, and understanding of quality control
3.3 How well education on statistical-mindedness and methods has taken
3.4 Understanding of effects
3.5 Education at affiliates (especially subcontractors, suppliers, and sales companies)
3.6 QCC activities
3.7 Suggestion system for improvements and its workings
4. **Collecting and Using Information**
4.1 Collection of outside information
4.2 Diffusion of information among divisions
4.3 Speed of information transmittal (e.g., use of computers)
4.4 Information processing and statistical analysis
5. **Analysis**
5.1 Selection of priority problems and themes
5.2 Correctness of analytical methods
5.3 Use of statistical methods
5.4 Relating to company's technology
5.5 Quality and process analysis
5.6 Use of results of analysis
5.7 Constructive use of suggestions
6. **Standardisation**
6.1 System of standards
6.2 Method of setting and revising of standards
6.3 Actual setting and revising of standards
6.4 Standards themselves
6.5 Use of statistical methods
6.6 Accumulated technical expertise
6.7 Use of standards
7. **Control**
7.1 System for controlling quality, cost, and production volume
7.2 Control points and control items
7.3 Use of statistical means (e.g., control charts) and statistical thinking
7.4 Contribution by QCCs

7.5 Actual control activities
7.6 State of control
8. Quality Assurance
8.1 New product development methods (e.g., quality analysis & deployment, reliability & design studies)
8.2 Safety and product liability prevention
8.3 Process design, analysis, control, and improvement
8.4 Process capability
8.5 Measurement and inspection
8.6 Equipment, subcontracting, purchasing, and servicing control
8.7 Quality assurance system and diagnosis
8.8 Use of statistical methods
8.9 Quality assessment and audit
8.10 Actual state of quality assurance
9. Effect
9.1 Measuring effect
9.2 Tangible benefits such as quality, service, delivery scheduling, cost, profit, safety, and environment
9.3 Intangible benefits
9.4 Convergence between expected and actual effect
10. Planning for the Future
10.1 Understanding of present conditions and specificity
10.2 Policies for correcting defects
10.3 Plans for promoting TQM
10.4 Relation to long-term plans

The Deming Prize Checklist reveals the important TQM areas ranging from the policy, organisation, training, information collection and analysis, standardisation, quality control and assurance, to planning for the next cycle of TQM. The process ensures that organisations will focus on all TQM areas for continuous improvement, from strategies to tactics and operations. One particular area worth discussing is the 'standardisation'. In Japan, standardisation does not just confined to conformance to national or international standards. The Japanese promote heavily the use of 'company standardisation'. In other words, if a process is proven to be effective, it should be standardised and used throughout the organisation. Then quality can be guaranteed despite the actual operations being in other parts of the world.

USA's Malcolm Baldrige National Quality Award (MBNQA)

The MBNQA is the *American response to the prestigious Japanese Deming Prize.* The Award was established by the US Congress in 1987 in remembrance of the

Chapter 9: TQM Kitemarks

popular Secretary of Commerce who died accidentally. The original idea was created by Admiral Frank Collins when he saw a half page advertisement in an American newspaper by a Japanese company. It said, "The most important name in quality control in Japan is America."

Baldrige is not a 'standard' but an annual competition like the Deming Prize. The Award is designed to recognise US companies that excel in quality achievement and quality management. The Award is designed to promote the following:
- an awareness of quality;
- understanding of the requirements of quality;
- sharing of information on successful strategies and the benefits derived from implementation.

The Baldrige criteria fit extremely well with the Deming philosophy of quality [Sallis, 1993]. However, unlike ISO 9000 there is a strong emphasis on the non-procedural aspects of quality such as leadership, human resource management, including employee well-being and morale, and customer satisfaction. This is necessary because Baldrige is a TQM Award which focus on continuous quality improvement through the employees in the organisation. Therefore, there need to be clear identifications on the deployment of human resources to achieve TQM. The analysis of the results of the quality improvement process are an important element. Applications are examined against a list of criteria which is regularly updated. The 1992 Examination Criteria had seven main parts as shown in Figure 9.1.

Item	Malcolm Baldrige Award Requirements (1992)	Max. Value	Evalu -ation
1	**Leadership (90 points)**		
1.1	Senior Executive Leadership	45	
1.2	Management for Quality	25	
1.3	Public Responsibility	20	
2	**Information and Analysis (80 points)**		
2.1	Scope and Management of Quality and Performance	15	
2.2	Competitive Comparisons and Benchmarks	25	
2.3	Analysis and Uses of Company-Level Data	40	
3	**Strategic Quality Planning (60 points)**		
3.1	Strategic Quality and Co. Performance Planning Process	35	
3.2	Quality and Performance Plans	25	

4	**Human Resource Development and Management (150 points)**		
4.1	Human Resource Management	20	
4.2	Employee Involvement	40	
4.3	Employee Education and Training	40	
4.4	Employee Performance and Recognition	25	
4.5	Employee Well-Being and Morale	25	
5	**Management of Process Quality (140 points)**		
5.1	Design and Introduction of Quality Products and Services	40	
5.2	Process Management - Product and Service Production and Delivery Processes	35	
5.3	Process Management - Business Processes and Support Services	30	
5.4	Supplier Quality	20	
5.5	Quality Assessment	15	
6	**Quality and Operational Results (180 points)**		
6.1	Product and Service	75	
6.2	Company Operational Results	45	
6.3	Business Process and Support Service Results	25	
6.4	Supplier Quality Results	35	
7	**Customer Focus and Satisfaction (300 points)**		
7.1	Customer relationship Management	65	
7.2	Commitment to Customers	15	
7.3	Customer Satisfaction Determination	35	
7.4	Customer Satisfaction Results	75	
7.5	Customer Satisfaction Comparison	75	
7.6	Future Requirements and Expectations of Customers	35	
	TOTAL	1,000	

Figure 9.1 MBNQA Examination Criteria (for 1992)

An example of how a company can win the MBNQA is shown in Mini-Case 9.1.

Mini-Case 9.1:

A 1991 MBNQA Winner: Zytec Corporation [Ross, 1993]

Zytec designs and manufactures electronic power supplies and repairs power supplies and monitors. It sells through direct sales force to customers in the US. However, since most customers are large multinational companies, many of Zytec's power supplies are exported around the world. Their competitors are from Far East and European companies as well as approximately 400 US companies. Zytec is currently the fastest growing US electronic power supply company, and the largest power supply repair company in North America. Since the beginning of the business in 1984, Zytec has used the quality and reliability of its products and services as the key strategy to differentiate from the competition. The following summary of Zytec's 1991 applications for the Award outlines the scope and success of its strategies in accordance with the MBNQA Checklist.

1 Leadership
1.1 Focus: Quality, Service and Value
1.2 TQM mission to Japan in 1988: Management By Planning (MBP)
1.3 'Non-customer visit day' on SPC, JIT, Deming, & world-class manufacturing

2 Information and Analysis
2.1 Customer-related data: delivery time, warranty repair, customer satisfaction levels, and Mean Time Between Failure analysis.
2.2 Follows a 6-step benchmarking process: plan, multi-research, observe, analyse, adapt, improve.
2.3 Benchmarking with Xerox, Motorola, Federal Express, IBM, DEC, Sony, HP, and Kodak.

3 Strategic Quality Planning
3.1 Gather data --> Long Range Strategic Planning cross-functional teams --> Departmental MBP teams.
3.2 Chief Executive reviews each team's objectives & action plans. Successes are commended. Problems are discussed.

4 Human Resource Development and Management
4.1 Guided by Deming's 14 points and Long Range Strategic Planning.
4.2 Self-managed work groups with broad authority.
4.3 Average 72 hours of internal quality-related training on SPC, QCC, JIT, Service America, etc.
4.4 Survey employees 7 times to gauge effectiveness of the Deming's Points.
4.5 Can spend up to US$1000 to resolve a customer complaint or make process changes with the agreement of only one other person.

5 Management of Process Quality
5.1 Four phases: Initiation, design initiation, sample delivery, pre-production and process certification.

5.2 Integrated Process Quality Control System including process flow analysis, SPC and design of experiment.
5.3 Management By Planning process involves all employees by empowering them to identify and implement Kaizen.
5.4 6-sigma 96% on-time delivery, and 25-day lead time.
5.5 Supplier selection, receiving inspection, supplier certification (at Parts-Per-Million levels)

6 **Quality and Operational Results**
6.1 Out-of-box quality improves from 99% (88) to 99.7% (90). Warranty costs reduced by 48%.
6.2 Product costs cut by 35%.
6.3 Reduce manufacturing cycle time by 26%. Inventory turn at world leadership level. More standard parts.
6.4 All improved significantly.

7 **Customer Focus and Satisfaction**
7.1 Sharing results with customers and listening carefully to their perceptions. All employees receive customer relationship training.
7.2 Every customer has been profiled and retention rate is high, with a growing customer base.
7.3 All customer complaints are managed by the Vice President of Marketing and solved.
7.4 Nine sources: sales representatives, account managers, quality reporting, executive phone calls, executive customer visits, employee customer visits, customer visits to Zytec, and customer satisfaction surveys.
7.5 Trends show continuous improvement. Original equipment manufacturing revenue grew 64% more than the industrial average. Adverse indicators have declined.
7.6 Zytec remains the sole source supplier for 18 of its 20 newly built customers.

■■■

The European Quality Award (EQA)

The EQA was launched during the 1991 European Foundation for Quality Management (EFQM) meeting in Paris. The Foundation is a new organisation formed in 1988 by 14 leading Western European companies. It is now made up of over 200 pan-European organisations who aim to stimulate and assist European companies in their development of total quality. The European Commission is playing a role in the development of EQA. The aim of the EFQM and of the Award is specifically geared to encourage the development of TQM. The Award aims *to recognise organisations who are paying exceptional attention to total quality, and to encourage others to follow their example.* The European Quality Award is not a standard but a competitive award like the Japanese Deming Prize and the US Malcolm Baldrige Award. It is a single annual award presented to the

Chapter 9: TQM Kitemarks

most successful TQM practices in Western Europe. In addition there are a number of European Quality Prizes (second best to be awarded) for companies who have demonstrated excellence in their management of quality.

In developing the European model of Corporate Excellence, the EFQM wanted to ensure that it not only reflected current thinking on best practice measures but also included some ideas of what a world class organisation would be measuring and achieving in the future. To reach a consensus, EQA Steering Committee was established to arrive at the model and the weightings within it. The result was a framework which had built in the strengths of its forerunners but added improvements with a distinctly European dimension (see framework in Figure 9.2).

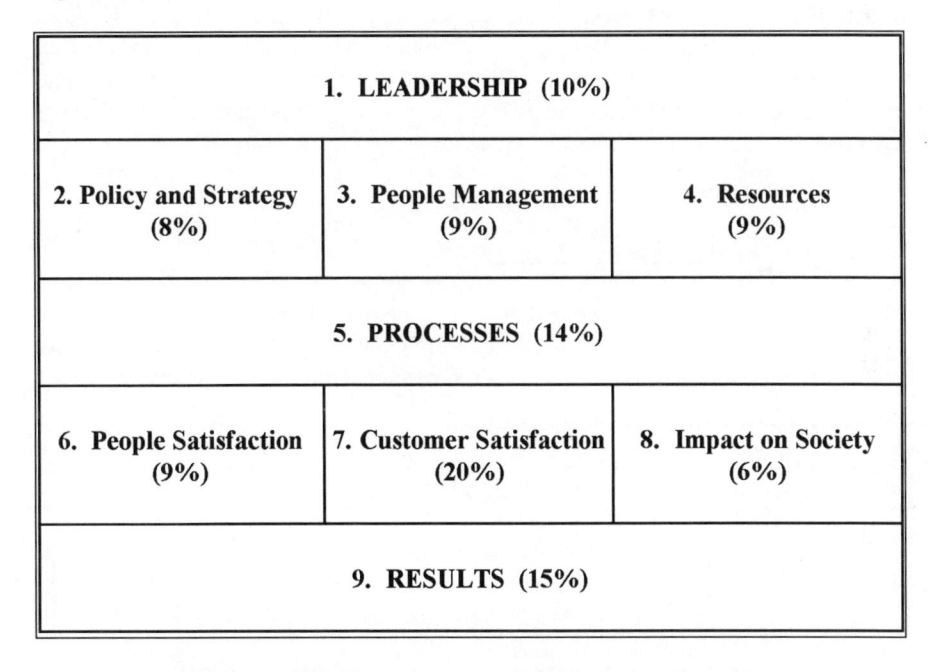

1. LEADERSHIP (10%)		
2. Policy and Strategy (8%)	3. People Management (9%)	4. Resources (9%)
5. PROCESSES (14%)		
6. People Satisfaction (9%)	7. Customer Satisfaction (20%)	8. Impact on Society (6%)
9. RESULTS (15%)		

Figure 9.2 The European/UK Quality Award Framework
(1-5 are the Enabling Factors; 6-9 are the Result Factors)

The 9-point Framework displayed in Figure 9.1 and detailed in **Annex 9.2** explain the content of the European Quality Award in greater details. The framework is prepared in the form of a checklist so that it can be used for evaluation readily.

The aim of the assessment methodology is to give great transparency and credibility to the award system. The process starts with the submission of detailed written reports by the applicant (of up to 75 pages). The applicant should address all the criteria in detail in the report. The assessors who score the report support their scoring by detailed comments. The scoring differences between the assessors are settled by a protocol.

The top scorers after this first round are subjected to a site visit of about 5 people for between 1 to 5 days, to verify and clarify the original submission. Again, there is a detailed code of conduct for the assessors during the site visit. They evaluate how the actual situation matches with the facts stated in the submitted report. The leaders of the assessment teams then report back to a panel of Jurors who make the final decisions. All applicants, whether subjected to a site visit or not, receive a written feedback from the EFQM.

The emphasis of the European Quality Award is: 'Are you doing the right things?' Compared with MBNQA, the EQA considers some areas as more important: impact on society, resource utilisation and business results. As the Award is new it is too early to assess its impact. The intention is to provide an award in the European Community of comparable status to the Japanese Deming Prize and the US Baldrige Award. The problem with it and other awards is that with only a few annual winners it is likely that its impact will be limited to some elite organisations. In effect, an important aim of the Award is to encourage organisations to learn from the achievement by the Award winners.

European Quality Award Results

The first European Quality Award was presented in 1992 to Rank Xerox Ltd. Each year, the jurors also selected some applicants to receive European Quality 'Prizes'. Figure 9.3 shows a summary of the winners since the EQA was launched.

Year	EQA Winner	Prize Awarded	Finalists
1992	Rank Xerox Ltd	• BOC Ltd., Special Gases • Industrias de Ubierna SA - UBISA • Milliken European Division	
1993	Milliken European Division	• ICL Manufacturing Division	• Cablelettra Spa (Italy) • Varian-TEM Ltd.
1994 (October)	D2D Ltd.	• IBM SEMEA (South Europe, Middle East and Africa) • Ericsson, Spain	• SCEMM, France • Texas Instruments Europe

Figure 9.3 EQA Winners, Prize Holders and Finalists

Chapter 9: TQM Kitemarks 235

The UK Quality Award

In the UK, the British Quality Foundation started the British Quality Award in 1990, with its own criteria checklist (see Mini-Case 9.2 & 9.3). In view of the popularity and conformity to the European Quality Award, the British Quality Foundation decided to start a new award called the UK Quality Award in 1994 (see also Figure 9.2). This award process replicates that of the EFQM using exactly the same model and assessment process.

1994 - November (1st round)

Five finalists were shortlisted for the UK Quality Award:
- Rover Group
- TNT Express (UK) Ltd
- BT Northern Ireland
- Avis Rent-A-Car Ltd.
- ICL Customer Service Division

Prime Minister John Major presented the first UK Quality Awards to two companies: **Rover Group** and **TNT Express (UK) Ltd** which have an equal status (see Mini-Case 9.3 & 9.5).

■■■

Mini-Case 9.2:

Girobank Wins the British Quality Award in 1991 [Henderson, 1992]

*Girobank was established in 1968 as a basic money transmission service and in two decades has developed into a major **clearing bank** with an account base of over two million personal and corporate customers and nearly 600,000 credit card holders. The bank employs some 6000 staff in 18 locations in the UK and each week handles over 8 million transactions and 250,000 customer contacts.*

Since 1987, Girobank has been involved in implementing a TQM programme. The strategy has focused on three principal activities -- quality improvement, customer care and quality assurance. The initiative started in the bank's processing area, which is regarded as the bank's factory or back office, and was progressively introduced into other areas of the bank, including accounts management which interfaces directly with the external customer.

The critical elements of the Girobank approach to TQM are a triple focus on:

*1. A **quality improvement programme** which concentrates on a broad range of improvement activities embracing:*

- the identification, elimination and prevention of errors, rework and wastage via a systematic audit process;
- the design control of all processes, products and services;
- the development of mutually beneficial partnerships with all suppliers;
- a shift in 'culture' values so that the whole of the workforce become active and effective 'change agents' in meeting business goals.

2. **A customer care programme** which focuses on problems perceived principally by the external customer but also embraces internal customer relationships and directs action towards activities which affect these customer perceptions, including:
- market research to identify customer needs, priorities and current perceptions of service achievement;
- staff attitude research, for it is their basic interaction with customers that will primarily determine customer perceptions;
- an effective complaints-handling system supported by a high degree of customer responsiveness;
- effective customer care training, particularly in the areas of product knowledge and the 'charm school' courtesies;
- effective customer communication and product presentation;
- ensuring that the resourcing and technological aspects of the customer response infrastructure are fit for purpose.

3. **A quality assurance programme** which sets standards for performance and implements a quality management system (ISO 9000) which ensures compliance to these standards. This system should ensure that all the benefits gained from the previous two processes are held and assured against decay.

Some of the benefits achieved so far are covered below:
1. Errors by keyboard operators have been reduced by 52% to 0.008%.
2. Post Office errors have been reduced by 65% to 0.06%.
3. Customer complaints have been reduced by 25%.
4. Inventory has been reduced by 38%.
5. Savings from quality improvement projects exceed £6 million.
6. Savings achieved from implementing staff suggestions total £2.6 million.
7. Girobank is a BSI Registered Firm.
8. Girobank is a British Quality Award Winner in 1991.
9. Customer perception on 'Quality of Service' has consistently improved over the past three years against the national measurement.

Mini-Case 9.3:

Short Brothers Plc Wins the British Quality Award in 1992   [McBride, 1993]

Short Brothers Plc *(Shorts)* is the largest industrial employer in Northern Ireland with some 8,500 personnel. The company has been engaged in the aviation business for over 90 years and its activities include the design and production of civil and military aircraft, the design and manufacture of major components for other aerospace corporations and the design and production of close air defence weapon systems.

Shorts launched a total quality programme in 1987; this signalled a period of unrivalled change within the company. Now, six years further on, Shorts can corroborate over £64 million worth of financial benefit to the organisation, can highlight a 50% involvement from their employees and in recent years have won a National Training Award (1989) for their total quality training, a commendation in the Northern Ireland Quality Award (1991) and both the Northern Ireland and British Quality Awards in 1992. The most significant was the establishment of a Total Quality Centre and associated staff, to facilitate the programme and the changes required within the company.

The role of the Centre was very clear:
1. to develop a total quality strategy for the company to achieve its business objective;
2. to develop initiatives which would enable the Divisions to support the Quality Councils and Functional Quality Teams;
3. to act as a focal point for total quality education and training;
4. to research and provide expertise on a range of quality improvement tools and techniques; and
5. to co-ordinate publicity for total quality activities.

The key strategic issues which have been identified as those which will improve the business are: Employee Involvement, Customer Focus, Measurement and Benchmarking. These four key elements form part of an overall continuous improvement strategy based on the European Quality Framework for Total Quality. Each member of the Total Quality Centre staff is assigned a responsibility for maintaining an awareness or working knowledge or an expertise on one of the following areas:
- Statistical Process Control
- Quality Circles / Natural Teams
- Taguchi Methodology
- Suggestion Schemes
- Quality Costs

- Benchmarking
- Customer Focus
- Quality Function Deployment

As a further means of performing research and acquiring new and relevant information, a member of the Total Quality Centre staff has been seconded to the University of Ulster as the Short Senior Fellow in Quality Management. This two-year post allows the job-holder to perform research identified as being key to the Shorts total quality strategy. The post also offers Shorts long-term benefits as the total quality teaching modules introduced on various degree courses at the University ensure that future employees of the company understand the concepts and practicalities of continuous improvement.

■■■

■■■

Mini-Case 9.4:

Rover Group Wins the First UK Quality Award   [Hakes, 1995]

Business results shown by **Rover Group**, *with sales of £4.3 billion in 1993, were related to its Total Quality Improvement programme started in mid-1987. While the European motor industry has been in a record post-war recession, Rover was the only car company to increase its sales in Europe -- up 8% in 1993 and 23% in the first half of 1994. Corresponding world-wide sales increases were 9% and 16%. Growth in productivity over five years was evidenced by average revenue per employee rising from £86,400 in 1989 to £126,000 in 1994.*

A wide range of methods has been used by Rover to track customer satisfaction with its products and dealer networks, with significant increases reported year by year since 1990 and confirmed by external surveys. Product quality improvements have led to major reductions in warranty costs, while customer confidence is now backed by the opportunity to return or exchange new vehicles within 30 days.

Rover's UK workforce stands at 34,000. Since 1986 Rover has conducted opinion surveys every other year among its employees, with findings acted upon through a structured feedback process. In the 1994 survey, 92% of people said they were proud to work for Rover, compared with 69% in 1990.

■■■

Mini-Case 9.5:

TNT Express (UK) Ltd. Wins the First UK Quality Award [Hakes, 1995]

TNT started its Total Quality initiative in 1989. Profit increases have been achieved every year since, with a 90% rise over the most recent five year period, matched by an increase in margins from 5.3% to 9.7%. TNT has proved the first choice carrier among UK businesses each year since 1991. Its growth has exceeded growth in the total market by factors of three in 1991, four in 1992 and six in 1993. The customer base has expanded from 62,000 in 1989 to 105,000 in 1993.

Other improvements over five years are expressed in on-time deliveries -- up from 94% to 97.8%, and unit costs per consignment -- down from £13.68 to £10.97. Lost time accidents are down by 48% since 1990.

In the latest attitude survey, 91% of the workforce of 6,600 responded that the organisation was committed to developing people. In 1993-4, TNT's £1.3m 'Expressing Excellence' programme provided structured customer care training for more than 4,000 employees. The firm won the Motor Transport Industry Training Awards in 1990 and 1993.

Why are TQM Kitemarks Useful?

"Self assessment, which every company ought to do anyway, is extremely revealing."
      -- Clive Jeanes, Managing Director of the European Division of
                            Milliken, 1993 EQA Winner.

TQM is being adopted by companies around the world. It is a key element to the competitiveness of a company that intends to win in the marketplace. As a testament to its importance, national and local governments sponsor many awards and prizes. The main reason that award processes such as the Deming Award, MBNQA and EQA are gaining acceptance is that *their underlying score quality values and principles (as stipulated in the requirements of a particular award) are commonly perceived as being beneficial and relevant to the development of most organisations* in today's climate of competitive global markets, fast paced technological innovation, and rapidly-changing working practices.

Awards and prizes are given to the organisation that achieves control over a comprehensive quality infrastructure, and show results from their TQM effort.

They all offer a framework against which a company is evaluated. The framework represents the basic systems and processes that support an organisation's quality initiative. It is sufficiently general to allow organisations reflect their own unique situations.

An analysis of the prizes is important to all interested in managing the quality organisation. An organisation can be introduced to the rigors of achieving high levels of quality by following one of these frameworks. The benefits of achieving a TQM kitemark are far reaching. For example, Masaaki [1986] has identified the benefits in acquiring the Deming Prize as follows:

1. **Tangible Benefits**
- Increased market share
- Increased sales volume
- Increased production volume
- Successful development of new products
- Shortening of product development time
- Development of new markets
- Improved quality
- Fewer complaints
- Reduced defect costs
- Fewer processes
- More employee suggestions
- Fewer industrial accidents

2. **Intangible Benefits**
- Increased involvement in management
- Increased quality consciousness and problem consciousness
- Better communications, both horizontally and vertically
- Improved quality of work
- Improved human relations
- Improved information feedback
- Improved management skills
- Permeation of market in concept
- Clear delineation between responsibility and authority
- More confidence in new product development
- Conversion to goal-oriented thinking
- Improved standardisation
- More active use of statistical quality control

The above benefits are also applicable to other kitemark achievers. Moreover, although not every contesting organisation can win the kitemark, or all the above listed benefits, a partial achievement is also a step towards TQM.

How to Achieve TQM Kitemarks

"Quality is an endless journey: like walking towards the horizon - no matter how far you walk, it does not change where the horizon is."

-- *Bernard Fournier, Managing Director of Rank Xerox, 1992 EQA Winner.*

It is clear that for each of the awards there are criteria to be met, and evaluations and visits to be conducted before the final selection of winners. On the other hand, ISO 9004-4 can be used as a guideline for quality improvement, an essential part of all the TQM kitemarks. Implementation of TQM kitemarks consists of two stages:

Stage 1: To decide on the most appropriate kitemark to achieve, and
Stage 2: To implement TQM according to the requirements effectively.

The first stage requires the study and evaluation of the requirements for the kitemark. A key task then is to select a framework for self-assessment. Deming Prizes are almost exclusively won by Japanese firms. The EQA and MBNQA are both excellent and robust models [Hakes, 1994]. If your organisation is not linked to Japan, USA or Europe, you can still choose one of these kitemarks as your TQM framework, or you can use your own national quality award (if such exists).

Most likely, your choice will be driven by the location or the parentage of your organisation. If your organisation is based in the USA or has a strong USA parent organisation then the MBNQA will probably be the most appropriate. If your organisations is based in Europe the EQA or its national derivatives (such as the UK Quality Award) will be the most appropriate model as it will additionally allow for direct comparisons and benchmarking with other European based organisations. Both of these models are comparable in the scoring and in most cases a 500 point out of 1000 score on the EQA model will equate to a similar score on the MBNQA model. In the final decision, geographic location will probably be the determinant of the model to be used.

After deciding on which of the kitemarks to adopt, the next stage is to implement the kitemark requirements effectively. The procedures for implementing TQM kitemarks are very similar to that of ISO 9000 (Chapter 7). However, the scope is broader than that of ISO 9000. Apart from the *quality assurance system*, TQM kitemark implementation should focus on *quality improvement* (as expressed in ISO 9004-4) and *customer care* as well. The best way to learn about kitemark implementation is from the example of a successful company (see Mini-Cases 9.1 to 9.5).

Chapter Summary:

- ► TQM kitemarks give important messages to customers, actual and potential, that the organisation takes quality seriously, and that its policies and practices conform to national and international standards of quality.
- ► The content of ISO 9004-4 includes the fundamentals for Quality Improvement (QI) such as: managing QI, methodology for QI and supporting tools.
- ► Japan's Deming Prize is the earliest and best-known prize for outstanding TQM organisations.
- ► USA's Malcolm Baldrige National Quality Award is the American equivalent of the prestigious Japanese Deming Prize.
- ► The European and UK Quality Awards aims to recognise organisations who are paying exceptional attention to total quality, and to encourage others to follow their example.
- ► TQM kitemarks are useful because their underlying score quality values and principles (as stipulated in the requirements of a particular award) are commonly perceived as being beneficial and relevant to the development of most organisations.
- ► Achieving TQM kitemarks entails the decision on the most appropriate kitemark to adopt, and the effective implementation according to its requirements.

Hands-on Exercise:

1. Consider the ISO 9004-4. Highlight the Quality Improvement parts which are relevant to achieving TQM for your own organisation.

2. Make a decision for your organisation to go for one of the several TQM kitemarks discussed in this Chapter. Justify your choice in terms of the suitability for your organisation.

3. Advise your Chief Executive on how you can manage the implementation of the kitemark of your choice in 2. above in your organisation within three years.

Annex 9.1 : ISO 9004-4:1993 Outline
(Section 4 to 7 only)

4. **Fundamental Concepts of Quality Improvement (QI) :**
   4.1 Principles of QI
   4.2 Environment for QI
      4.2.1 Management responsibility and leadership
      4.2.2 Values, attitudes and behaviour
      4.2.3 QI goals
      4.2.4 Communications and teamwork
      4.2.5 Recognition
      4.2.6 Education and training
   4.3 Quality losses
5. **Managing for QI :**
   5.1 Organising for QI
   5.2 Planning for QI
   5.3 Measuring QI
   5.4 Reviewing QI activities
6. **Methodology for QI :**
   6.1 Involving the whole organisation
   6.2 Initiating QI projects or activities
   6.3 Investigating possible areas for QI
   6.4 Establishing cause-and-effect relationships
   6.5 Taking preventive or corrective actions
   6.6 Confirming the improvement
   6.7 Sustaining the gains
   6.8 Continuing the improvement
7. **Supporting Tools and Techniques :**
   7.1 Tools for numerical data (see details in the following table)
   7.2 Tools for non-numerical data (see details in the following table)
   7.3 Training in applying tools and techniques

Sub-clause	Tools and Techniques	Applications
A.1*	Data-collection form	To gather data systematically to obtain a clear picture of the facts.
Tools and techniques for non-numerical data		
A.2	Affinity diagram	To organise into groupings a large number of ideas, opinions or concerns about a particular topic.
A.3	Benchmarking	To compare a process against those of recognised leaders to identify opportunities for quality improvement.
A.4	Brainstorming	To identify possible solutions to problems and potential opportunities for quality improvement.
A.5*	Cause-and-effect diagram	To analyse and communicate cause-and-effect diagram relationships. To facilitate problem solving from symptom to cause to solution.
A.6*	Flowchart	To describe an existing process. To design a new process.
A.7	Tree diagram	To show the relationships between a topic and its component elements.
Tools and techniques for numerical data		
A.8*	Control chart	Diagnosis: to evaluate process stability. Control: to determine when a process needs to be adjusted and when it needs to be left as is. Confirmation: to confirm an improvement to a process.
A.9*	Histogram	To display the pattern of variation of data. To communicate visually information about process behaviour. To make decisions about where to focus improvement efforts.
A.10*	Pareto diagram	To display, in order of importance, the contribution of each item to the total effect. To rank improvement opportunities.
A.11*	Scatter diagram	To discover and confirm relationships between two associated sets of data. To confirm anticipated relationships between two associated sets of data.

* Same as the 7 QCC Tools (see Chapter 6)

Annex 9.2 : The European Quality Award Checklist

The objective of the questionnaire is to guide the managers responsible for the function to assess quality related issues, so that the results can be analysed and compared.

Firm: _____   Date: _____

Respondent: _____   Post: _____

(*) This is a 11-point scale, i.e., 0=Lowest ---> 5=Average ---> 10=Highest.

No.	Evaluation Criteria: Assessment of ...	Value(*)
1	**Leadership (10%)** -- This criterion looks at how managers drive the organisation towards the achievement of organisational excellence.	
1.1	Visible Involvement -- How managers are actively and visibly involved in driving towards the achievement of organisational excellence based on the concepts and principles of Total Quality.	
1.2	Culture -- How managers develop and maintain an environment / culture consistent with continuous improvement and corporate excellence.	
1.3	Recognition -- How individuals and teams are recognised by managers.	
1.4	Support -- How managers support improvement activities with appropriate resources and assistance.	
1.5	Customers and Suppliers -- How managers are involved in improvement and other activities with customers and suppliers.	
1.6	Promotion -- How managers promote continuous improvement and quality values outside the organisation.	
2	**Policy and Strategy (8%)** -- This criterion looks at your Mission, Vision, Values and Strategies and how you plan to achieve them.	
2.1	Quality Values and Concepts -- How the concepts of total quality are reflected in your policies and strategies.	

2.2	Relevant Information -- How you use all sources of information, relevant to organisational excellence, as inputs to your policies, strategies and plans.
2.3	Business Plan Deployment -- How your plans are effectively produced and aligned to the achievement of business goals, policies and strategies.
2.4	Policy and Strategy Communication -- How policies, strategies and plans are communicated.
2.5	Regular updating and improvement of policy and strategy -- How you review and improve your strategies, policies and plans.
3	**People Management (9%)** -- This criterion looks at how your organisation releases the full potential of its people.
3.1	Continuous Improvement Practices -- How people management processes are aligned to business needs and continuously improved.
3.2	Training, Recruitment and Career Progression -- How effective use is made of recruitment, training and career progression processes to develop skills and capabilities.
3.3	Targets for People and Teams -- How team and individual targets and objectives are agreed and how these are reviewed.
3.4	Involvement and Empowerment -- How everyone is empowered and encouraged to become involved in continuous improvement activities.
3.5	Communication -- How communication is made two-way and effective.
4	**Resources (9%)** -- This criterion looks at the ways your organisation manages its key resources in support of its policy and strategy.
4.1	Financial -- How you manage your financial resources effectively.
4.2	Information -- How you manage your information.
4.3	Material -- How you manage your material resources, fixed assets and supplies/suppliers.
4.4	Technology -- How you identify and manage appropriate, alternate and emergent methods and technologies.

5	**Process** (14%) -- This criterion looks at how you identify, manage, review and, where appropriate, revise your organisation's processes.	
5.1	Identifying Critical Processes -- How you identify and define the key processes within your organisation.	
5.2	Managing Processes -- How you manage ALL the processes within your organisation.	
5.3	Measures, Targets and Reviews -- How you use all relevant information, including measurement and feedback, to review processes and set improvement targets.	
5.4	Innovation and Creativity -- How you stimulate innovation and creativity in process improvement.	
5.5	Process Change -- How you implement and evaluate process changes.	
6	**People Satisfaction** (9%) -- This criterion looks at the organisation's achievement in relation to the satisfaction of its people.	
6.1	Direct Results -- Perception measures and results, i.e., judgement by the people/employees of what the organisation is achieving in relation to the satisfaction of its people.	
6.2	Indirect Results -- Predicting, leading and influencing measures and results, i.e., judgement by the organisation of factors that are likely to influence employee satisfaction.	
7	**Customer Satisfaction** (20%) -- This criterion looks at the organisation's achievement in relation to the satisfaction of its external customers	
7.1	Direct Results -- Perception measures and results, i.e., judgement by the customers of the organisation's products, services and customer relationships.	
7.2	Indirect Results -- Predicting, leading and influencing measures and results, i.e., judgement by the organisation relating to customer satisfaction.	
8	**Impact on Society** (6%) -- This criterion involves assessing what the organisation is achieving in satisfying the needs and expectations of the community at large. This includes views on the organisation's approach to the quality of life, the environment and to the preservation of global resources.	

8.1	Direct Results -- Perception measures and results, i.e., judgement by community at large of the organisation's impact on society.	
8.2	Indirect Results -- Predicting, leading and influencing measures and results, i.e., judgement by the organisation of factors that are likely to influence 'Impact on Society'.	
9	**Results (15%)** -- This criterion involves an assessment of what the organisation is achieving, in relation to its planned business performance, in satisfying the needs of everyone with an interest in the organisation and in achieving its planned business service objectives.	
9.1	Direct Results -- Financial, 'Bottom Line' results that indicate an organisation's successes.	
9.2	Indirect Results -- Non financial key efficiency and effectiveness measures and the results of the key processes identified in Criteria 4 and 5 that are vital indicators of an organisation's current and continuing success.	
	TOTAL WEIGHED SCORE	

Chapter 10
TQMEX in the Real World

Topics:

- *Validation of the TQMEX Model*
- *The UK survey results and analysis*
- *The UK/Japan survey results and analysis*
- *Significance of ISO 9000 in TQMEX implementation*
- *How to implement TQMEX*

```
5-S
 ↓
BPR
 ↓
QCCs
 ↓
ISO
 ↓
TPM
 ↓
TQM
```

Validation of the TQMEX Model

One important feature of TQMEX is the ability to *offer a step-by-step procedure in achieving TQM*. Furthermore, each individual step can be used in its own right, and its results can be assessed separately. This has a great advantage because companies can make a choice where to focus their effort, and even go back to the previous steps if they have not yet done so. The model is simple and flexible.

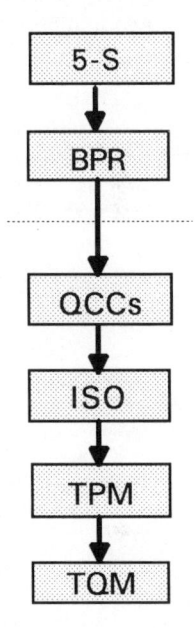

In an attempt to prove that all the TQMEX elements are sound management methods and good quality practices, and to convince the reader of their effectiveness, I have recently conducted an intensive questionnaire based survey of UK and Japanese organisations which have been practising TQMEX. The following discussion will outline the findings.

Sampling

In selecting the samples for this survey, the main criteria was that the companies chosen had already adopted some good quality practices. Since ISO 9000 gives the formal certification and guarantees that the company is quality conscious, this was the yardstick in the selection process. This survey had two aims: the first was to assess the TQM practices contained in the TQMEX Model of the UK firms (**Stage 1**) and the second one was to draw a comparison of the TQM practices between leading UK and Japanese firms (**Stage 2**).

The sample addresses were obtained from the *"DTI QA Register 1994"* because it is the most comprehensive directory of quality assurance registered firms in the UK. The intention was to have a balanced sample of companies from both sectors of industry regardless of their size and ownership.

Stage 1 of this survey was conducted in August 1994 on a random sample of 1800 registered firms. About 10% valid replies were obtained from both the manufacturing and services industries: 51 manufacturing firms with average employee number of 968 and 110 services firms with average employee number of 283. Figure 10.1 contains the industrial classifications of the respondents.

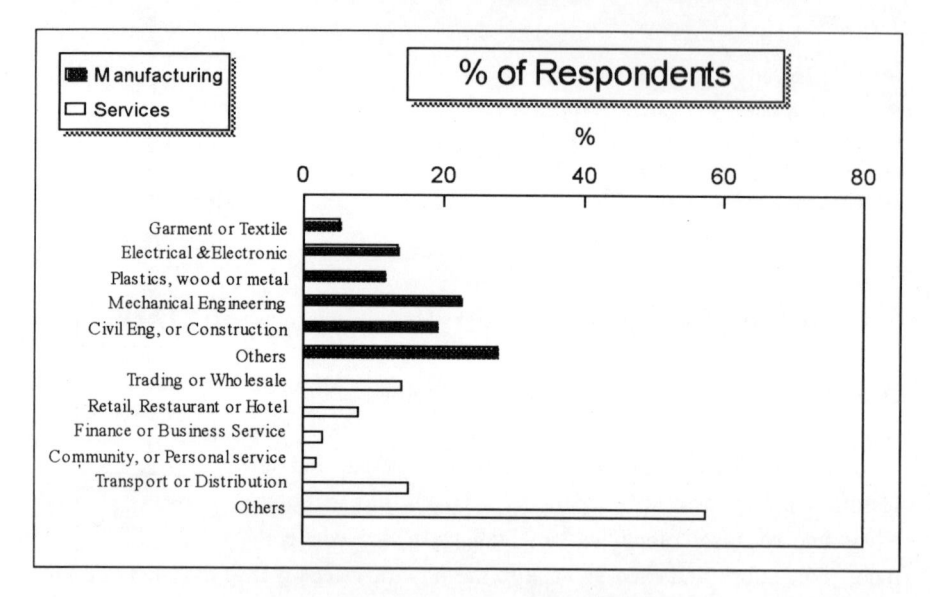

Figure 10.1 Percentage of Respondents

Stage 2 of this survey was conducted in January 1995 on 200 leading UK and 200 leading Japanese firms. They are selected from the *"Times 1000 -- 1995"* and *"Extel Financial Asia Pacific Handbook 1992"*. About 8% of replies were valid. Since the main objective of the second stage was to do a comparative study, rather than establishing statistical significance, even this low feedback was considered adequate. From the feedback questionnaires received, 18 leading companies of approximately the same size (in terms of number of employees) were selected from each country for the analysis.

The Questionnaire

The questionnaire consisted of two parts. **Part 1** was designed to obtain background information about the surveyed company, including:
- Ownership of the company
- Existing workforce

Chapter 10: TQMEX in the Real World

- Management perception of existing system
- Methods used for improving quality
- Quality training frequency

Part 2 consisted of questions related to the seven elements of the TQMEX Model:
- 5-S Practices
- Business Process Re-engineering (BPR)
- Quality Control Circles (QCC)
- ISO 9000 quality management system (QMS)
- Total Preventive Maintenance (TPM)
- Total Quality Management (TQM)
- Opinion about QMS

Note: *Seventy questions were asked in total. A 7-point scale was used for most questions. The last section of Part 2 (QMS) had a form of three open-ended questions. The questionnaires were answered by quality managers of surveyed companies.*

Survey Results

Stage 1 Results: The UK Companies Survey

The following analysis will shed the light on the extent to which the UK companies are familiar with the six elements of the Model and how far they have gone in implementing them.

Japanese 5-S Practices

Figure 10.2 reveals that over 80% of the surveyed companies have not come across the 5-S concept.

Figure 10.3 shows that each of the activities has a mean between 5 and 6. When the Analysis of Variance (**ANOVA**) statistical procedure was used to test the data, it shows that services sector has a significantly greater percentage in carrying out the 5-S when compared with manufacturing. One explanation is that the majority of services have to serve their customers face to face and therefore environmental condition is important in attracting more customers. Another result from the ANOVA is that both manufacturing and services industries consider organisation as more important than cleaning and neatness.

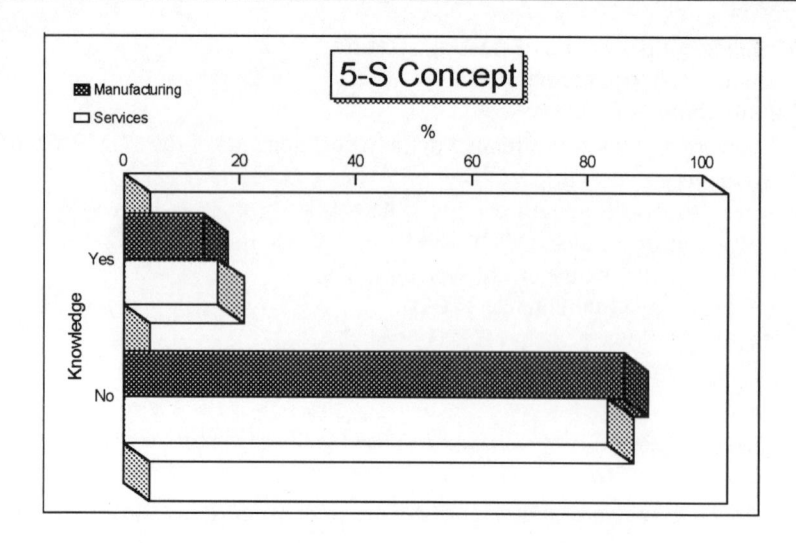

Figure 10.2 The 5-S Concept

From these findings, it is apparent that the majority of the UK companies practise the 5-S to some extent without realising it. It would therefore be worthwhile to formalise the 5-S practice so that more companies can implement it as a matter of course.

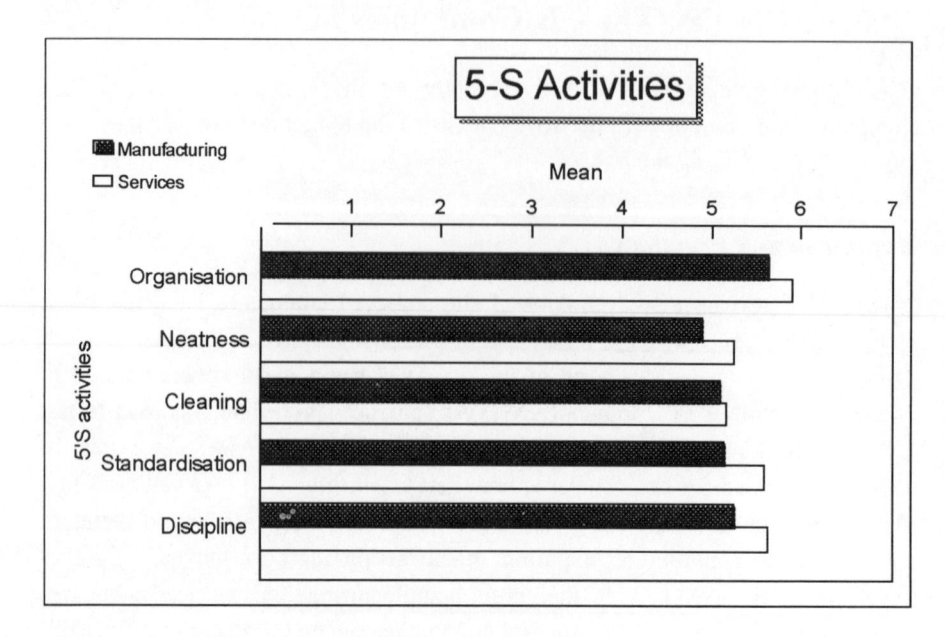

Figure 10.3 The 5-S Activities

Business Process Re-engineering

There is a strong link between quality and business functions such as marketing, contracting and purchasing. In any company, the marketing and purchasing departments have the closest contact with the customers at both ends of the supply chain. Meeting the needs of the customer is what TQM is all about. Therefore it is vital for a company to review and balance their emphasis on quality control for the input (purchasing) and output (marketing) functions as well.

Figure 10.4 summarises the feedback on quality control applied to marketing, production and purchasing functions. The ANOVA results revealed that companies have put more emphasis on production/operations as the key areas for quality control, whilst paying less attention to other areas, marketing in particular. The survey finding on BPR has proved the fact the total quality marketing should require more attention by companies in the UK.

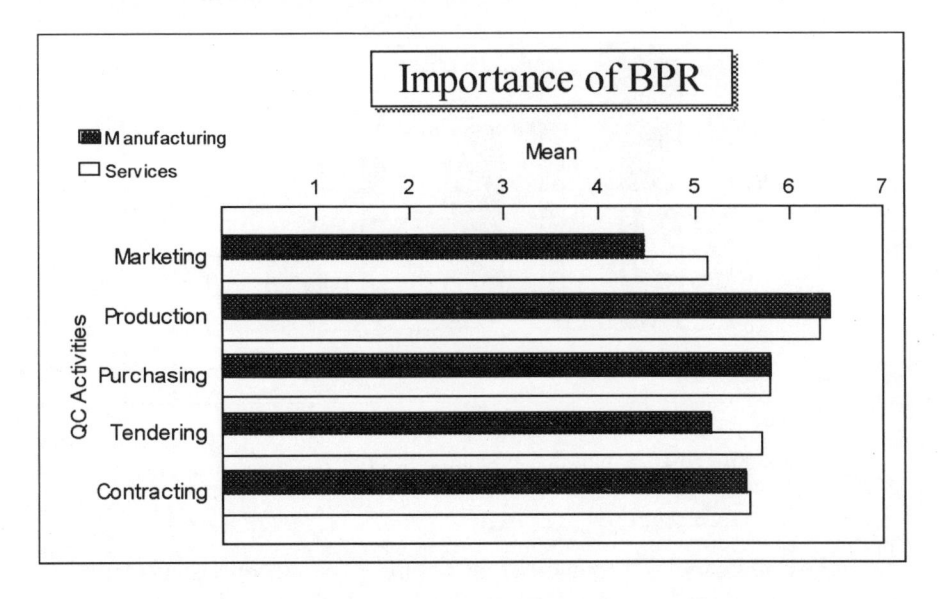

Figure 10.4 Importance of Business Process Re-engineering

Quality Control Circle

It is the Japanese experience that 95% of the problems in the workshop can be solved with simple Quality Control methods such as the 7 tools of QC (refer to Chapter 6). These tools will help QCCs to do brain-storming systematically and analyse the problems critically. Most problems can be solved without pain through logical thinking and experience.

The result of the survey (Figure 10.5) shows that 50% of manufacturing and 35% of the services companies have been practising QCCs. In half of those companies, QCCs did not make any improvement suggestions during the previous year. As for the other half, the average number of suggestions are 103 (in manufacturing sector) and 250 (in services sector) per company per year. This is very low compared with the Japanese standard -- more than 20 suggestions per circle per annum. This is probably due to lack of management motivation and staff training on the use of QCC tools.

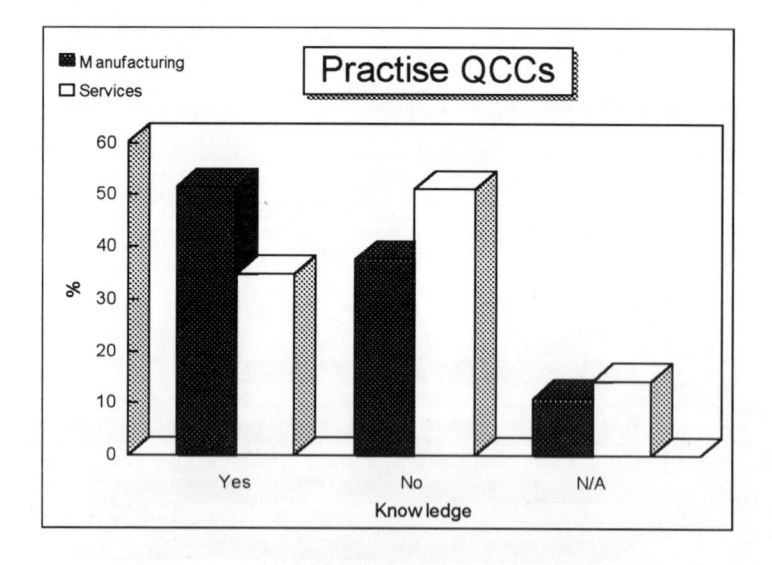

Figure 10.5 Number of companies practising QCCs

The survey feedback is summarised in Figure 10.6. There is a significant difference in appreciation of QCCs between the manufacturing and services sectors. The service sector views QCCs as more important because they believe that QCCs can help them to improve their service quality directly. Nevertheless, manufacturing sector sees QCC as an important tool, though somewhat less than the service sector. The ANOVA also reveals that 'improving quality' is seen as more important than 'building cheerful environment'.

According to the General Manager of an ISO 9000 registered manufacturing company, without the participation of top and middle management, QCC activities cannot be sustained. To encourage participation of all departments, the

Chapter 10: TQMEX in the Real World

company has introduced a reward scheme for any circle that comes up with the best solutions.

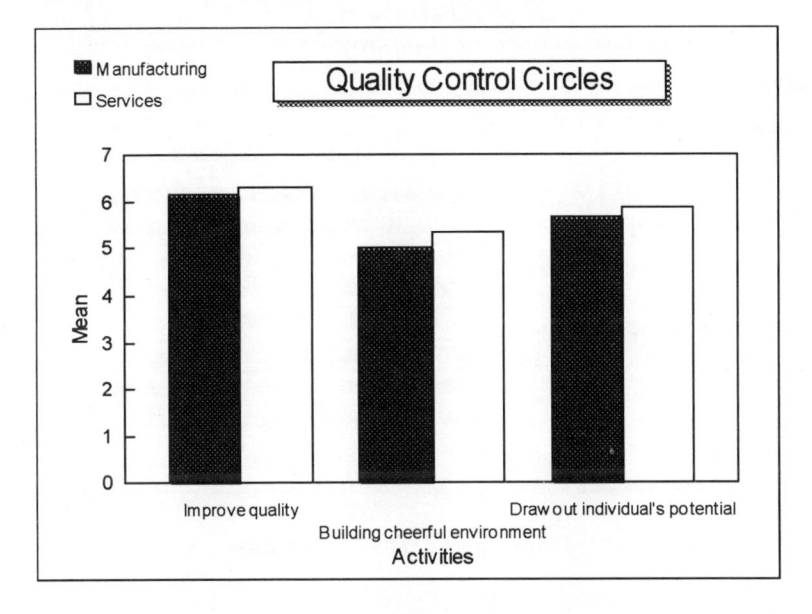

Figure 10.6 Quality Control Circles

Although useful to improve quality, QCCs are not widely practised in the UK. QCCs can bring about cultural change and motivate the staff to contribute to the improvement of their work. Therefore, they should deserve more attention from management and be promoted throughout the organisation.

ISO 9000 Quality Management System

Figure 10.7 shows the results of the survey on the effectiveness of ISO 9000 QMS. The results revealed that registered companies enjoyed more of the benefits of increasing quality awareness, attracting and satisfying customers **(soft benefits)** than the reduction of waste, time and cost **(hard benefits)**. An explanation for this is that companies often expect ISO 9000 to give them cost savings and other hard benefits within a short time. Unfortunately, this benefit is not immediately achievable. In order to measure the hard benefits, suitable measuring systems, such as progress measurement, error-cause analysis, waste measurement, etc., have to be built in. In practice, companies should put the objective of satisfying customers (soft benefits) as their first priority. Then, monetary (hard) benefits will follow as a matter of course.

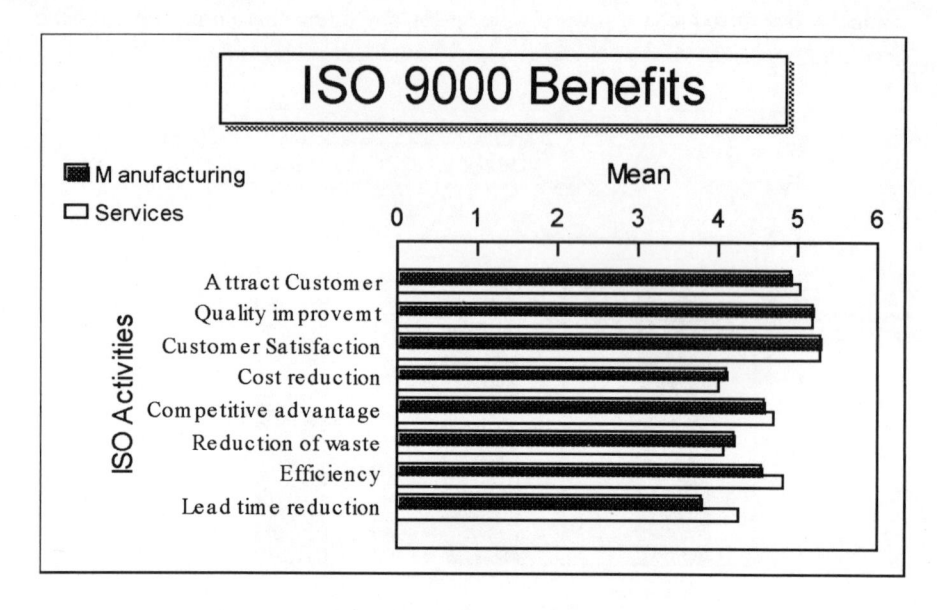

Figure 10.7 ISO 9000 Benefits

Total Productive Maintenance

Figure 10.8 shows that about 60--70% of the companies from both manufacturing and services sectors have not practised TPM. A likely reason for this is that companies often do not know what benefits TPM could bring to them. More TPM training is therefore necessary.

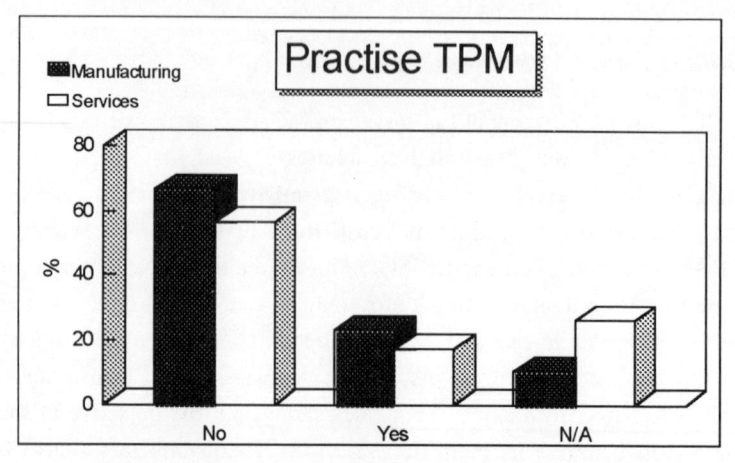

Figure 10.8 Number of companies practising TPM

Chapter 10: TQMEX in the Real World 257

Figure 10.9 shows that there is a significant difference in the use of TPM between manufacturing and services sectors. Manufacturing employs more TPM techniques than their services counterpart. This agrees with the fact that a lot more of the services firms do not know about TPM at all. The survey results show that on average, 70% of maintenance time is spent on emergencies (fire fighting), 15% on non-critical repairs and only 15% on planned preventative maintenance.

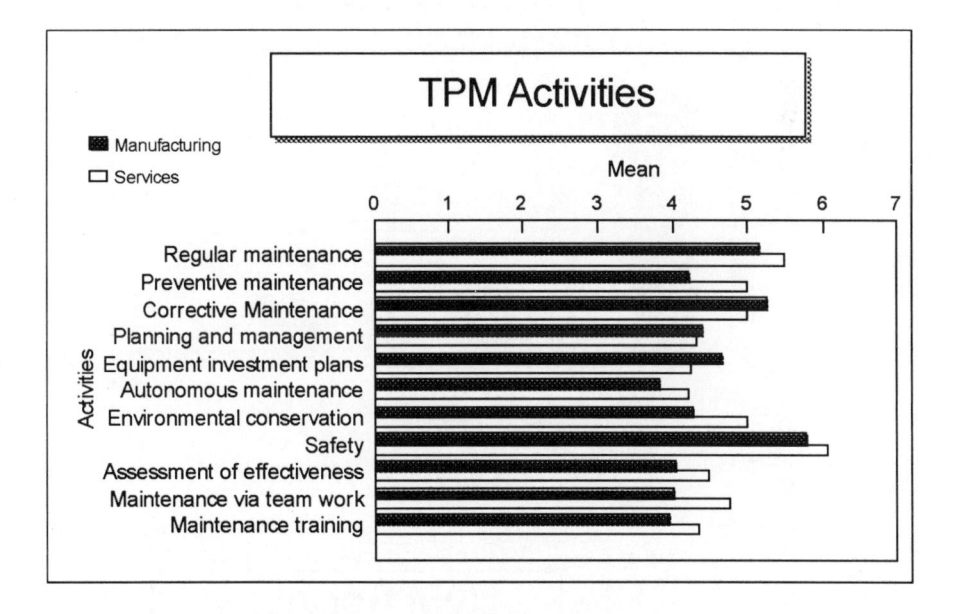

Figure 10.9 TPM Activities

It is evident that maintenance has been conducted in a casual manner without real effort in trying to incorporate all maintenance aspects. This is mainly due to the management's negative attitudes towards the role of maintenance in relation to business operations. This was confirmed by a survey conducted by the Department of Trade and Industry in the UK to establish best practices in relation to maintenance which came up with the following conclusions:
- Maintenance is not always considered at company executive level;
- Most companies seem to ignore the real cost of downtime in terms of lost sales opportunities
- Within manufacturing industry, only 3.7% of annual sales revenue is spent on maintenance of operations equipment;
- The survey has also concluded that with good maintenance management the above costs can be drastically reduced and plant availability increased to lead up to 30% increase in profitability.

Total Quality Management

Figure 10.10 shows that 70% of manufacturing and 50% of services recognised a need for TQM, apart from being registered to ISO 9000.

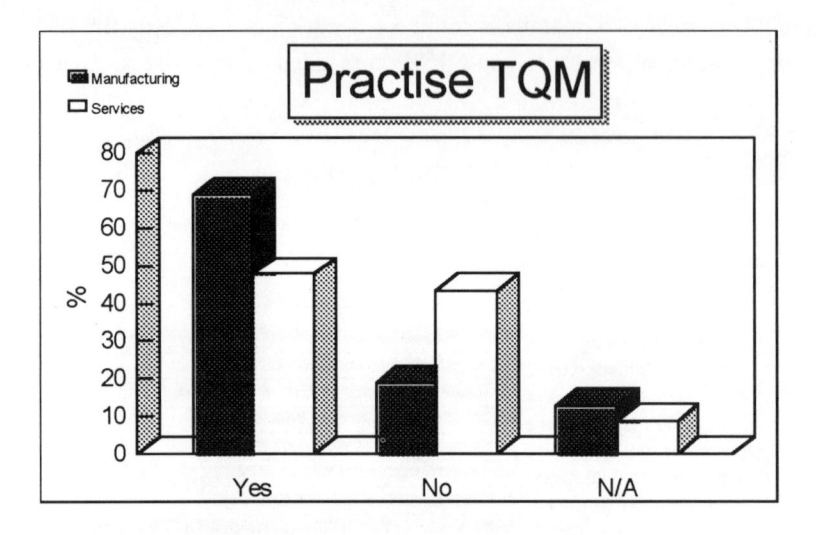

Figure 10.10 Number of companies practising TQM

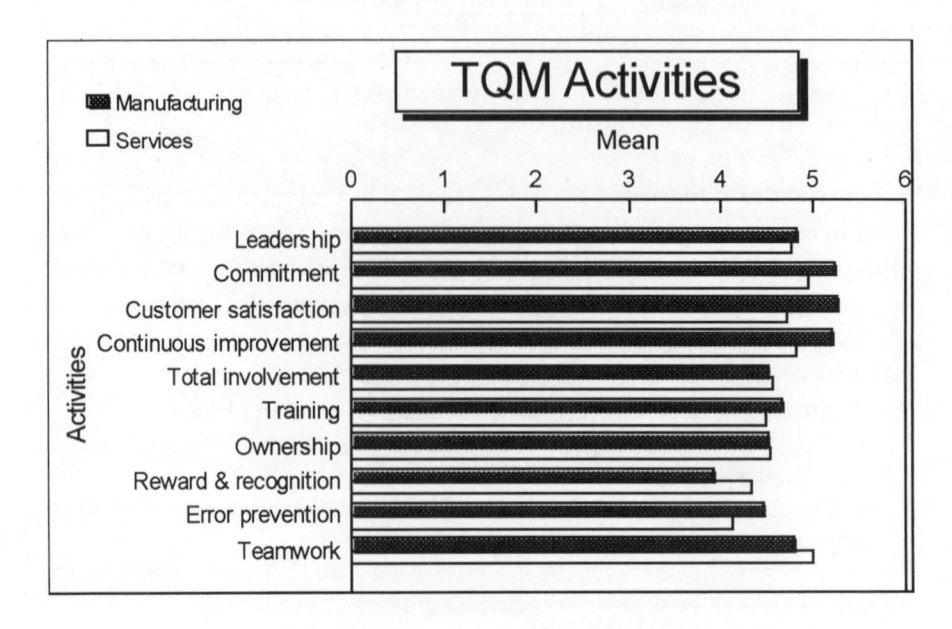

Figure 10.11 TQM Activities

Chapter 10: TQMEX in the Real World 259

Figure 10.11 shows that the mean score ranges from 4 to 5.3. The ANOVA test result shows that there is no significant difference between the activities, i.e. companies consider all TQM activities as equally important.

Open-ended Questions on Quality Management System

The final set of questions were open-ended and encouraged opinions and suggestions on **merits, drawbacks** and **possible improvements** to company's existing QMS. The comments provided some important conclusions and recommendations to be considered by firms and they are summarised as follows.

1. **Merits**
   - Some quality managers feel that their QMS provides them with better organisational structure, improves awareness of quality and commitment from top, attracts more customers, makes the company more customer oriented, provides more training, and improves image.
   - The main merits of the QMS employed are to identify actual responsibilities, provide control of procedures and product quality, improve understanding of customer--supplier relationship and provide a base for TQM
   - Some quality managers expect ISO 9000 to give them cost savings within a short time-period. About 10% claim quality costs reduction.

2. **Drawbacks**
   - Some quality managers have difficulties in dealing with continuous paperwork, and in maintaining the system which requires a lot of human resources.
   - Lack of senior management commitment and training.
   - Most companies are unable to give details of drawbacks as they are still in the early stage of implementing the QMS.

3. **Possible improvements (as quoted)**
   - Long term strategic thinking.
   - Greater emphasis on quality-related training to all levels.
   - Continuous improvement of equipment, working environment, attitudes and behaviour of people.
   - More emphasis on quality circles, better policy deployment, procedure simplification and greater ownership.
   - Use of Information Technology **(IT)** to reduce paper work and increase communication link between departments.
   - Setting up operator teams to look at improvement area in their work place.
   - Reduction in bureaucracy.
   - More management involvement on quality cost reduction and more identifiable financial returns to increase top management commitment.

- Intention to look closely at benchmarking, value engineering and performance appraisal.
- There are no "quick fixes". The quality culture is well embedded and fairly well thought out. Better communication between QA and other department would help. Closer integration with health, safety and environmental management systems.
- Development from product quality to service levels and to increase process control and further deployment of responsibilities to lowest levels possible.
- More customer feedback, elimination of inspection, more automation of process.

Most of the possible improvements quoted above can be achieved by adopting good practices based on the TQMEX model, as discussed in Chapters 4 to 8. A good marketing, production and purchasing control in modern day business described in the BPR chapter, relies heavily on the proper use of IT, including computer networks and effective software.

Summarising different views from various companies will bring some light to the merits and drawbacks of quality management systems. In order to achieve excellence, all the steps in the TQMEX model are important for the betterment of quality.

	Manufacturing: Mean of means	Services: Mean of means	Weighed: Mean of means
5-S	5.2	5.5	5.29
BPR	5.3	5.2	5.26
QCC	5.5	5.9	5.62
ISO	4.6	4.4	4.53
TPM	4.5	4.8	4.59
TQM	4.7	4.6	4.66
Overall	4.9	.5.0	4.93

Figure 10.12 Mean Scores for the 6 Groups of Questions

Figure 10.12 summarises the mean of means of the six groups of questions for both manufacturing and services industries. It is seen that the overall mean for both industries is on the high side on the 7-point scale (all means are above 4, the medium of the scale). This is expected as the samples are from registered

established their QMSs. Following the principle of continuous improvement, registered companies should aim at doing better than the means stipulated in Figure 10.12. The information from this Figure should also serve as a target for the unregistered companies aiming to adopt TQM principles.

When comparing the relative mean values of the 6 groups of questions, it is obvious that ISO 9000 QMS score is the lowest. This demonstrates that the companies see the ISO 9000 quality system as a passport to stay in business. On the other hand, the highest mean score is associated with QCCs. About half of the companies have practised QCCs and they strongly believe that QCCs are important for the company.

The second highest rating in the Figure is 5-S. Apparently, this is unexpected as the term '5-S' has not been widely used amongst the UK companies. However, quality companies are always concern about the 5-S constituents as a matter of course. The British are disciplined in organising their workplace in order to establish a conducive environment, the 5-S can be promoted easily and readily in the UK.

Summary of UK Survey Findings

The analysis of the survey results among 161 registered UK companies operating across a range of manufacturing and services industries have provided evidence for the step by step approach of TQMEX. Each of the 6 sets of questions have been analysed for UK firms. Every step of TQMEX is considered as important for achieving TQM. The respondents were also aware of most of the improvement tools such as BPR, QCC, ISO 9000 and TPM. 5-S needs to be promoted in order to maximise its benefits. All of the steps listed in the TQMEX Model have been rated very high by the companies. This proved that TQMEX is a workable and practical model for implementing TQM.

Comments from the three open-ended questions are also summarised. They further endorse the importance of TQM and the need for correct approach. For the TQMEX model to be applicable, training must be conducted continuously at every stage of the TQM programme. It is important to keep quality systems short and simple, in order for TQM to work effectively. Both the manufacturing and services industries must have greater management commitment to long-term strategic planning, training and quality improvement. Finally, TQM will pay off the resources invested in it and guarantee a successful future for the company.

Stage 2 Results: The UK and Japanese Leading Companies Survey

For the comparative study of companies from the UK and Japan, the *5-S, TPM and TQM* will be discussed as typical constituents of the TQMEX Model.

Japanese 5-S Practices

Figure 10.13 reveals that over 80% of the surveyed companies have not come across the 5-S concept (this is the same set of result as in Figure 10.2). The leading firms' replies are more encouraging. About 80% of the Japanese companies have practised the 5-S as compared to 40% in the UK. The ANOVA has revealed that this difference is statistically significant.

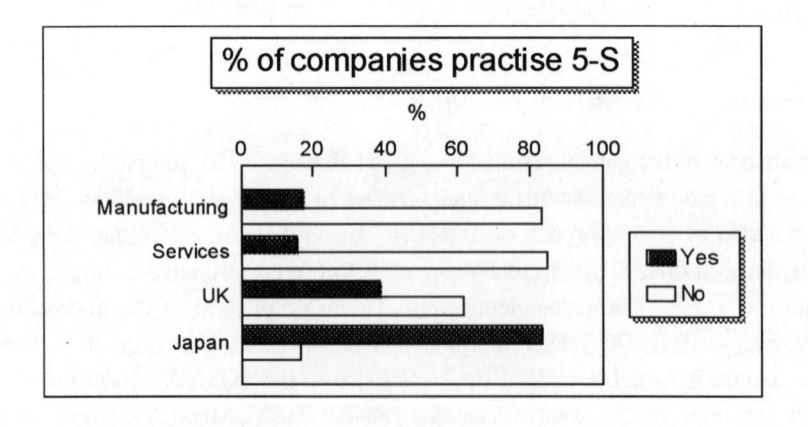

Figure 10.13 Number of Companies Practising the 5-S

In comparing UK and Japanese firms, Figure 10.14 shows that each of the activities has a mean between 5 and 6. There is no significant difference in practising the 5-S activities between the firms from the two countries.

Chapter 10: TQMEX in the Real World

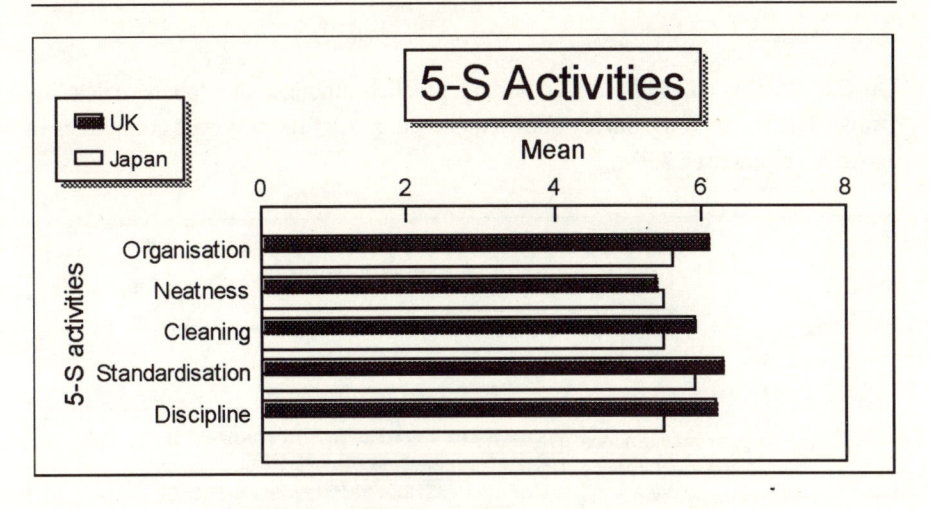

Figure 10.14 5-S Activities (UK vs. Japan)

Total Productive Maintenance

Figure 10.15 shows percentage of both manufacturing and services sectors practising TPM, as resulted from the Stage 1 survey (see Figure 10.8). Only about 10% of the UK leading companies surveyed in Stage 2 practise TPM. A likely reason for this is that often companies do not know what benefits TPM could bring to them. The questionnaire answers also suggest that, on average, 70% of maintenance time is spent on emergencies (fire fighting), 15% on non-critical repairs and only 15% on planned preventative maintenance.

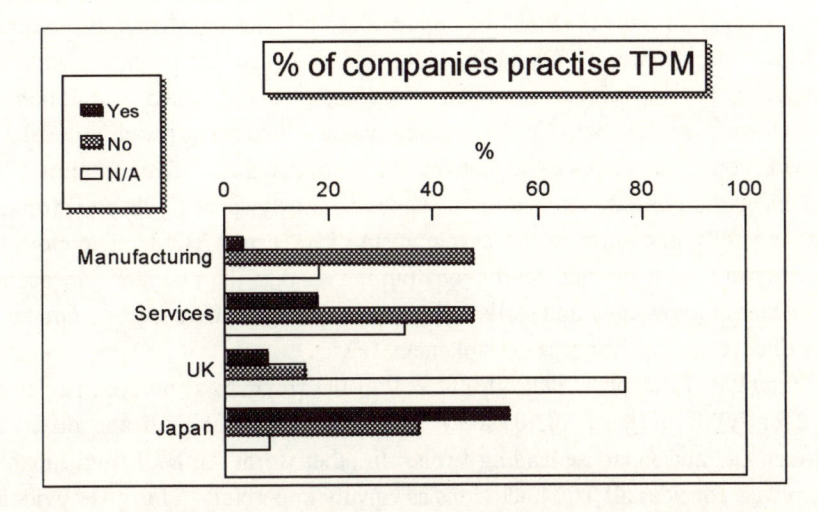

Figure 10.15 Number of companies practising TPM

On the other hand, TPM is a common practice amongst the Japanese leading firms. The result of the survey shows that over 50% of the surveyed companies in Japan have practised TPM.

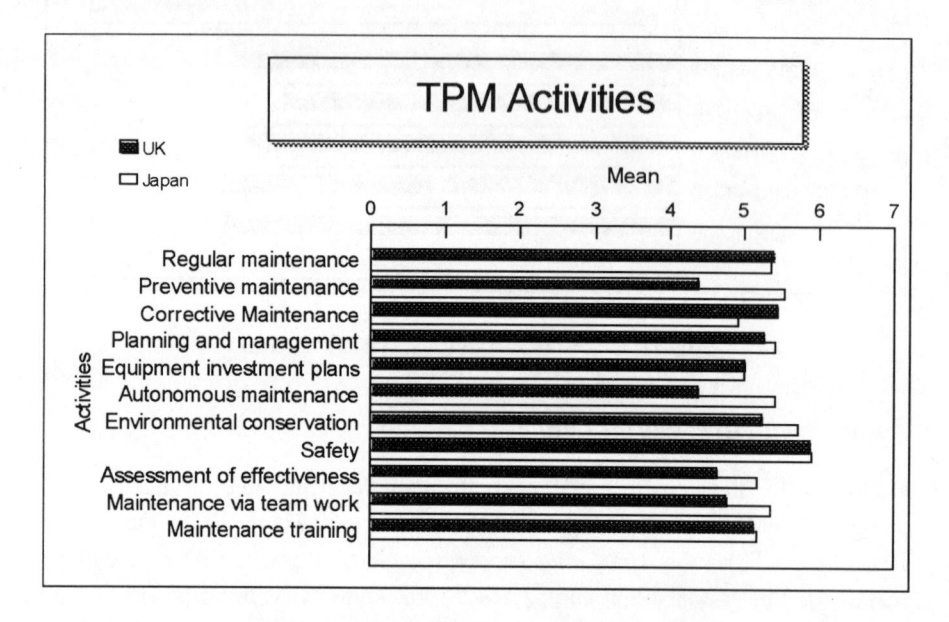

Figure 10.16 TPM Activities (UK vs. Japan)

The results of the survey shown in Figure 10.16 give no significant difference in TPM activities between UK and Japanese leading firms. However, the Japanese companies have a higher mean in autonomous maintenance. Autonomous maintenance is a key point in TPM development. Increasing people's interest in carrying out maintenance cannot be done overnight. Time required for implementation of autonomous maintenance is largely spent to change attitudes. This is vitally important in the development of TPM and TQM. Therefore top management must provide on-the-job training so that the operator can acquire maintenance knowledge and skills as he moves through different jobs. Gradually he will develop a willingness to implement TPM.

Figure 10.17 displays that 50--60% of all the companies surveyed perceive a need for TQM. Figure 10.18 shows that there is no significant difference between UK and Japanese leading firms. In other words, both UK and Japanese companies consider all TQM activities as equally important. This gives evidence that TQM has now become universal, not just in Japanese firms.

Chapter 10: TQMEX in the Real World

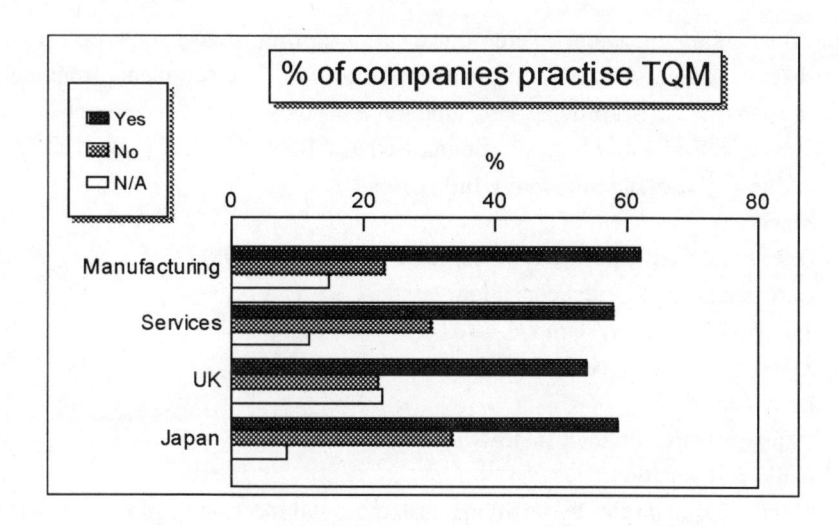

Figure 10.17 Number of companies practising TQM

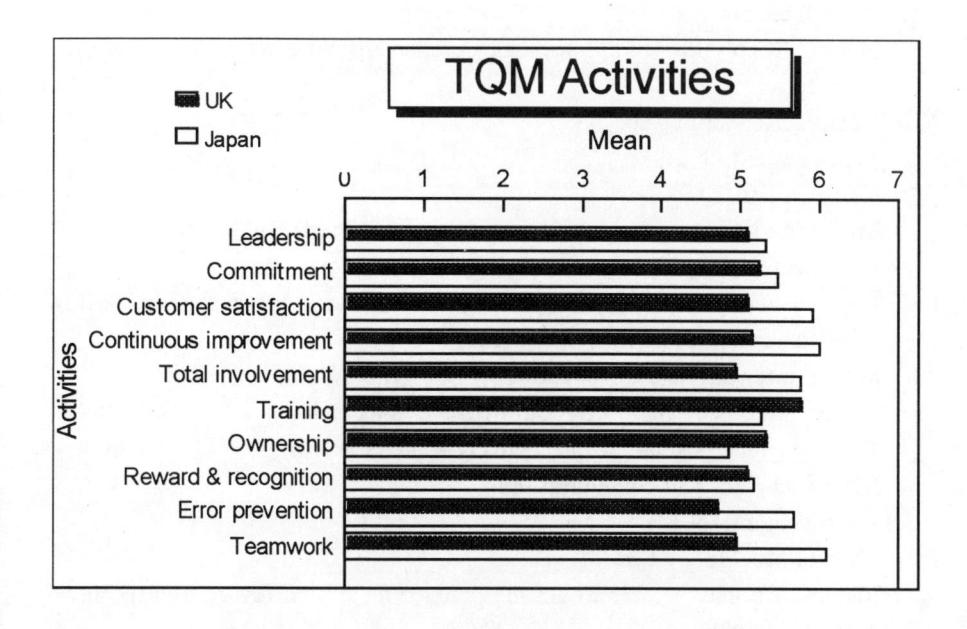

Figure 10.18 TQM Activities (UK vs. Japan)

Open-ended Questions on Quality Management System

The final three questions were aimed at acquiring suggestions on merits, drawbacks and possible improvements to their QMS. The comments from some leading Japanese firms provide some important conclusions and recommendations for firms to consider and they are summarised as follows.

1. **Ishikawajima-Harima Heavy Industries**
   Merit: Improvement of company constitution.
   Drawback: All employees do not really understand TQM.
   Improvement: Correct education.

2. **Japan Technical System Co. Ltd.**
   Merit: Increase of company reliability.
   Drawback: N.A.
   Improvement: Product stability for software.

3. **Kaijo Corporation**
   Merit: Expand sales by satsifying customers and increase employees' morale.
   Drawback: Give an influence on daily work when on the way to establish perfect QMS.
   Improvement: N.A.

4. **Kirin Brewery Ltd.**
   Merit: Supply the most satisfactory products, saving cost and activate the organisation.
   Drawback: Nothing special.
   Improvement: N.A.

5. **Komatsu Ltd.**
   Merit: Increased concern about own performance and achievement.
   Drawback: More likely to start losing cross-functional activity.
   Improvement: Reporting system to top management and evaluation skill of top management about their subordinate.

6. **Medox International Corporation**
   Merit: Improve employees' knowledge (let them know what they can do, should do, and must do for the company and themselves).
   Drawback: Lack of communication.
   Improvement: N.A.

7. **Murata Machinery Ltd.**
   Merit: We may be able to get the practice for design review and technical documentation.
   Drawback: None.
   Improvement: We can get the uniformity for quality, but we may lose the individual superiority.

Chapter 10: TQMEX in the Real World

8. **NEC Corporation**
   Merit: Concentration of employee's consciousness towards company strategic targets in the competitive market.
   Drawback: Give the basic conditions to tender in the overseas market by ISO 9000 certification and keep the interest.
   Improvement: Involve the meta-Quality concepts such as safety, environment conservation, etc., and to deploy their targets to each division and subsidiary.

9. **Nippon Seiki Co. Ltd.**
   Merit: Improvement of efficiency by reducing failure rate, which results in cost reduction.
   Drawback: Weak organisation and lack of leadership.
   Improvement: Improvement of quality through the results of the preparation to become an ISO 9000 registered company.

10. **NKK Corporation**
    Merit: Employee's ownership, quality improvement.
    Drawback: Restructuring.
    Improvement: Systematic policy deployment and evaluation.

11. **OMIC**
    Merit: Top down leadership.
    Drawback: N.A.
    Improvement: Changes and improvements are ongoing.

12. **Sanyo Corporation**
    Merit: Consensus among senior executive. The spirit of being in the same boat. Promotion of improvement activity.
    Drawback: Indirect sector (white-collar staff) need to adopt TQM on a greater scale.
    Improvement: Emphasise on human factors, leadership by top managers, thorough customer orientation.

13. **Sato Kogyo Co. Ltd.**
    Merit: All the improving action on quality has been considered and carried out based on customer satisfaction.
    Drawback: The way to achieve customer satisfaction could not be seen clearly.
    Improvement: N.A.

14. **Tajima Junzo Ltd.**
    Merit: Shorten the delivery time.
    Drawback: Complicated organisation.
    Improvement: Comprehensive organisation in order to have co-operative teamwork and recognition.

15. **Tokyo Electric Power Company**

   Merit: Employees have worked with the concepts of market-in PDCA Cycle, more improvements achieved by QCCs.

   Drawback: Head Office was left behind in the improvement activities. The form of the activities rather than the content has become the objective.

   Improvement: Create and develop our own TQM system, need understanding among line manager. TQM should be implemented by line managers, not by TQM facilitators.

16. **Toyota Motor Corporation**

   Merit: All employees in their respective functions pledge to consider "Customer first", master basic methodologies of QC, and put them into practice, as a result it can improve corporate robustness.

   Drawback: Nothing for the time being.

   Improvement: TQM concepts have to be continued for the establishment of the corporate foundations under any circumstances from now on, although it is natural that the emphasis of TQM activities should be changed according to the environmental changes, for example in focus on environmental conservation, marketing strategy or employee satisfaction.

Summary of UK-Japan Survey Findings

The analysis of the survey results among 16 leading UK companies and 16 leading Japanese companies operating across a range of manufacturing and services industries have provided evidence for the similarity and differences between the two countries' approach to TQMEX.

About 80% of the Japanese companies have practised the 5-S compared with 40% in the UK. UK companies need to have better awareness of the applications of the 5-S to improve their working environment.

Only 10% of UK companies have practised TPM, compared with 50% of Japanese companies. UK companies have to put in a lot more effort to improve their productive maintenance by devoting more effort to improve their equipment and facilities.

Finally, both UK and Japanese leading companies consider all TQM activities as equally important. This gives evidence that TQM has now become universal, not just aimed for Japanese firms.

Comments from the three open-ended questions are also summarised. They further endorse the importance of TQM and the need for correct approach. When these set of responses are compared with the set from the Stage 1 survey, the most significant difference is that the Japanese emphasis is more on the education and training of staff to do kaizen spontaneously (see Mini-Case 10.1).

Chapter 10: TQMEX in the Real World

■■■

Mini-Case 10.1:

TQM Promotion by PHP Institute Inc., Tokyo   [Matsushita, 1988]

Matsushita Electric is the world's largest home electrical/electronic appliance manufacturer. It was founded by Konosuke Matsushita in 1918. Even back in 1932, he believed that the objective of manufacturing is to produce good quality products which can meet the needs of the customers at an affordable price, so that everyone can get it just like tap water.

Matsushita established the PHP (Peace and Happiness through Prosperity) Institute in 1946, just after the Japanese defeat in the Second World War. As the name implies, PHP is charged with the mission to "bring peace and happiness to the 20th century through prosperity". Matsushita used the idea of 'tap-water philosophy' because when the supply of goods is abundant, the society will become prosper and people become happier.

PHP continues to help the industries and society to improve the quality of work and life. In so doing, PHP publishes many of the books and videos which are considered important to achieve its mission. The publication is heavily focused on TQM. An analysis of their recent publications confirms once again that the TQMEX is an important model. Each of the elements of the Model has been published as a video seminar by PHP. Some of the titles are:

5-S: 5S: Five Steps to Shaping Up the Shop Floor
In-depth 5S - Questions, Answers, and Examples
5S for the Office
Visual Control Systems
BPR: Just-in-Time Case Studies
Techniques for Streamlining Production
QCC: Invitation to QC
Mastering the Tools of QC
The Power of the Suggestion System
ISO: Understanding the ISO 9000 Standards
TPM: This is TPM: Total Productive Maintenance
Zero Breakdowns
Zero Set-up Losses
TQM: You and Your Job: Having the Right Attitude
Teamwork on the Shop Floor: Key Points for Energising the Workplace

■■■

Significance of ISO 9000 in TQMEX implementation

Successful organisations have realised that continuous improvement is the way to sustain competitiveness and satisfy the changing needs of the various stakeholders (including external customers). Organisations need innovation and renewal, and TQMEX provides a framework for it. Another angle to look at the need for TQMEX is based on the popularity of ISO 9000 QMS.

ISO 9000 QMS has been implemented by over 40,000 firms world-wide and the number is growing at an unprecedented rate. This is what the Secretary-General of ISO called 'The ISO 9000 phenomena' [Eicher, 1992]. Accredited organisations have gone through a lot of hard work to keep ISO 9000 running. However, after acquiring the ISO 9000, many organisations are wondering what they should do next. There need to be new ideas to keep the organisation alive and growing.

It has been stated in Chapter 7 that ISO 9000 is a route to TQM. Since there are some controversial views, we need to consider four possible models of the relationship between ISO 9000 and TQM [Sallis, 1993].

Model 1: Sees ISO 9000 as the starting point for TQM. ISO 9000 tackles the procedural infrastructure which precedes the more difficult changes of culture and attitudes. Obtaining ISO 9000 provides the institution with kitemarked confidence to go forward to tackle the larger issues associated with TQM.

Model 2: Positions ISO 9000 at the heart of TQM. In this model, ISO 9000 holds TQM in place and provides it with a solid foundation for continuous improvement.

Model 3: Considers ISO 9000 as a minor role in TQM. ISO 9000 is seen as only one element in a more important venture. Its role is little more than a useful means of assuring the operational consistency of the organisation's procedures. In this model quality is delivered by the active participation of the human resources in improvement teams and not by paper-based procedures.

Model 4: Takes a negative view of ISO 9000 on TQM. In this model, ISO 9000 is considered as irrelevant to the pursuit of quality. ISO 9000 is viewed as a bureaucratic intrusion into the organisation's quality improvement. ISO 9000 has aroused some strong and hostile feelings. The concern is whether the extra workload and the need to work strictly to systems and procedures albeit internally generated, could damage staff morale and creativity.

The above four models are arranged in the descending order of preference for ISO 9000. However, we should not neglect the implication of Model 4. In other words, when we are developing the ISO 9000 system, we must make the quality manual short and simple, and above all useful to the organisation for quality improvement rather than as document bureaucracy. Models 1, 2 and 3 confirm

Chapter 10: TQMEX in the Real World 271

the value of ISO 9000 for TQM to a greater or lesser extent. The question to follow is: how can we convert an ISO 9000 system into a TQM system?

The Answer Lies in the TQMEX Model

If an organisation has already implemented ISO 9000, it should still go back to the first step, the 5-S practices, in order to develop a total quality environment. One task of the 5-S is to throw away the rubbish, including the obsolete documents and paperwork generated from the ISO 9000 system. The 5-S keeps on insisting to look into the root of the problem - why there are so much wasteful documents created in the first instance?

Then BPR will help to re-focus the business, making it more customer-oriented. This should then be built into the ISO 9000 system during management review meetings (Clause 4.1.3).

QCCs will contribute continuous improvement by mobilising everybody in quality initiatives. The 5-S provide good agenda for improvement. Furthermore, QCCs are good organisations to review the effectiveness of the ISO 9000 system and help communicate and understand the requirements of the Standard. This will lead to simplification of the quality manual.

TPM, when developed, will improve quality and productivity dramatically. This will be an important help towards TQM. If you walk into an efficient factory or office with conducive environment, you do not have to look at their ISO 9000 system before you can tell whether it is a quality organisation.

Finally, TQM is a process, not a destination. As Deming said, "We have to do it forever." When TQM is built upon 5-S, BPR, QCC, ISO and TPM, it will guarantee continuous improvement and customer satisfaction, no matter how demanding that could be.

Figure 10.19 lists some of the challenges that should be addressed by each unit within a firm. The improvements aimed at are classified into Musts and Wants. Musts are the criteria that have to be met by the firm for survival. Wants are the criteria that the firm needs to fulfill as competitive advantage. The effect of each of the TQMEX elements on improving the customer requirements are highlighted in the table. It can be seen that all of the six elements are essential for a firm to stay in business and remain competitive.

Item	IMPROVEMENTS (M = Must, W = Want)	5-S	BPR	QCC	ISO	TPM	TQM
M1	Reduce product/service lead times by ____%.		•	•		•	•
M2	Introduce new products/services at 2 times the present rate.	•	•				•
M3	Reduce product/service costs by ____%.		•	•	•	•	•
M4	Reduce overhead/support labour by ____%.	•	•				•
M5	Reduce customer complaints by ____%.	•	•	•	•		•
W1	Is your office/shopfloor clean, tidy, uncluttered?	•				•	•
W2	Are your processes and office equipment completely dependable and reliable?	•	•			•	•
W3	Are your design, inspection and other documentation clear, up-to-date and always used correctly?	•	•	•	•		•
W4	How much importance do you attach to developing your process improvement?			•	•	•	•
W5	Is your workforce totally flexible?	•		•	•		•
W6	Are you always achieving the shortest possible throughput times?	•	•			•	•
W7	Are you committed to a continuous improvement plan on all performance measurements?	•		•	•	•	•
W8	Are you actively committed to training, retraining and competence throughout the whole firm, including management?	•		•	•	•	•
W9	To what extent do you consider the subordinates as a source of ideas?	•		•		•	•
W10	Do you accept that continuous change will be the future of life in services and act on this principle?	•		•	•	•	•

- Where the TQMEX element will have a significant impact on the Musts and Wants.

Figure 10.19 The Impact of the TQMEX on Improvements to Organisation

How to Implement TQMEX

Firms can acquire the TQMEX Model through a series of training and implementation programmes. Figure 10.20 shows a rough-cut estimate of the time of the resource person who may be an experienced quality manager of the organisation or a quality consultant. The size of the organisation is assumed to be medium. The larger the organisation, the longer will be the time required for training.

The TQMEX Training & Implementation Programme is a proven approach to achieve TQM. It also provides an important opportunity for firms to improve themselves. Organisations which have gone through some stages of improvement can tailor the programme to their special needs.

BRIEF DESCRIPTION	**MAN-DAYS (*)**
A. A.1 Training/Consultancy Requirement Studies	5
A.2 Establish Quality Council	5
B. Conduct Training & Implementation on:-	
B.1 Housekeeping Through 5-S Practices	5
B.2 Business Process Re-engineering	5
B.3 Quality Data Collection & Seven QC Tools	5
B.4 QCC & Problem Solving	5
B.5 ISO 9000 Quality Manual Documentation	5
B.6 Internal Quality Audit	5
B.7 TPM Implementation	5
B.8 TQM Implementation	10
C. Documenting & Implementing ISO 9001/2 QMS	20

(*) *This is a rough estimate of the time of the resource person who may be an experienced quality manager of the organisation or a quality consultant.*

Figure 10.20 Suggested TQMEX Initial Training/Implementation Programme

The Implementation Cycle

Dr. Deming used to say, "TQM will last forever." Therefore it is important to understand that the TQMEX implementation should not be a programme in its

own right. It should be a continuous process, subject to improvement cycle as shown in Figure 10.21. The Deming Cycle should be followed at each stage of the TQMEX implementation. There should also be measurable inputs and outputs, so that the changes can be identified for further improvement. The following steps serve as the guidelines:
- Establish appropriate measures
- Define measurable inputs and outputs
- Define and document current procedures and working practice
- Identify problems that cause errors and reduce productivity
- Develop and implement corrective actions.
- Standardise the procedure
- Repeat activities for continuous improvement

and finally,
"I hear and I forget, I see and I remember, I do and I learn."
-- *Confucius, Chinese philosopher, 500BC.*

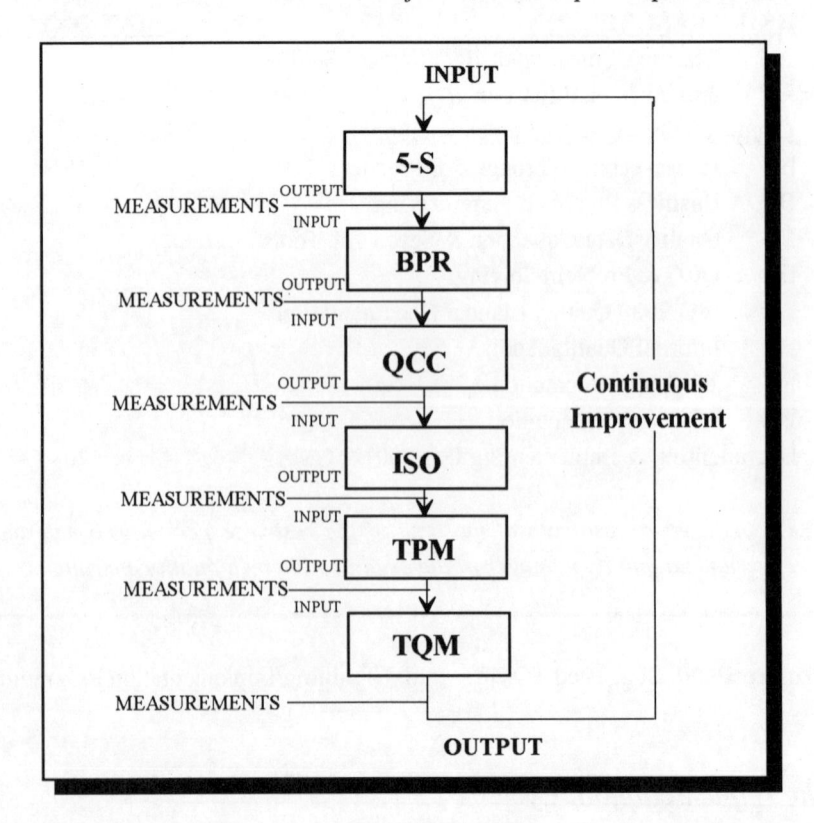

Figure 10.21 Implementation Cycle for TQMEX

Chapter 10: TQMEX in the Real World

Chapter Summary:

- *Validation of the TQMEX Model -- In order to justify the effectiveness of the Model and its elements, a two-stage survey was conducted. Stage 1 focused on the assessment of the UK ISO 9000 registered companies, while Stage 2 provided comparison between the UK and Japanese leading companies.*
- *Analysis of the Stage 1 results shows that all of the steps listed in the TQMEX Model have been rated very high by the UK companies involved in the survey. This proved that TQMEX is a workable and practical model for implementing TQM.*
- *Analysis of the Stage 2 results shows that the 5-S and TPM are more commonly practised in Japan than in the UK. All the companies, regardless the country, considered TQM important and the number of UK companies practising TQM is close to that of the Japanese.*
- *Opinion on quality management system -- All the companies involved in the survey further endorse the importance of TQM and the need for the correct approach. For the TQMEX model to be applicable, training must be conducted continuously at every stage of the TQM programme.*
- *Although the opinions about the relationship between ISO 9000 and TQM are divided, the phenomenal world-wide acceptance of ISO 9000 QMS must not be neglected. Having TQM as its ultimate goal, it offers a firm foundation for TQMEX implementation process.*
- *TQMEX can be implemented through a series of training and implementation programmes. Each stage of the implementation process should be based on the Deming Cycle.*

Hands-on Exercise:

1. Consider the TQMEX Model in the light of your own organisation's status quo in quality management. What are the steps you would take to implement TQMEX?

2. From the questionnaire survey results of the TQMEX, identify the survey finding which you can use to convince your Chief Executive that you have confidence to implement TQM successfully?

3. Based on Figure 10.19, re-define the MUSTs and WANTs for your own organisation and complete the table with •'s accordingly.

4. Draw up your TQMEX training and implementation plan.

REFERENCES

Abraham B. (1995, Jan). "The Last Word: Lesson of a Lifetime in Managing for Quality", *Quality World,* UK, pp.30-32.
Abramovitch I. (1994, Feb). "Beyond Kaizen", *Success,* Vol. 41, USA. pp.85-88.
APO (1986). *Productivity Management at Firm Level,* Asian Productivity Organisation, Tokyo.
Bajaria H.J. (1995). "Effective TQM Implementation: Critical Issues", *Proceedings of the 1st World Congress for TQM,* Chapman & Hall, UK, pp.128-136.
Ball P.L. (1992, Jun). "Strategy for Total Quality", *Quality Forum, Vol.18, No.2,* UK.
Brewin R.A. and Starkey M.W. (1995). "Applying Shewart's Statistical Charts for Improved Logistical Performance", *Proceedings of the 2nd International Symposium on Logistics,* Uni. of Nottingham, UK.
BSI (1987). *ISO 9000 / ISO 9000: 1987 A Positive Contribution to Better Business, an Executive guide,* London.
Bulled, J.W. (1987, November). "ISO 9000 - Quality Management Systems and Assessment", *General Engineer,* UK, pp.271-280.
Carlzon J. (1987). *Moments of Truth,* Harper & Row, New York.
CIM (1992, Oct). *Marketing,* Magazine of the Chartered Institute of Marketing, UK.
Crosby P.B. (1979). *Quality is Free: The Art of Making Quality Certain,* McGraw-Hill, New York.
Crosby P.B. (1984). *Quality without Tears: The Art of Hassle-Free Management,* McGraw-Hill, New York.
Crosby P.B. (1992). *Completeness,* Penguin Books, New York.
Dawson D. (1988, Dec 8). "Packaging in Marketing: Cracker Packs", *Marketing,* UK, pp.42.
Deming W.E. (1986). *Out of the Crisis,* Cambridge, MIT Centre for Advanced Engineering Study.
Deming W.E. (1989). *Foundation for Management of Quality in the Western World,* Perigee Books, USA.
Deming W.E. (1993). *The New Economics for Industry, Government, Education,* Massachusetts Institute of Technology, Center for Advanced Engineering Study, USA.
DoE (1982). *A Nation at Risk,* Department of Education, US Government Publisher.

Dignan W. (1995). "Business process re-engineering at the co-op Bank: improving personal customer service", *TQM Magazine, Vol.7, No.1*, MCB, UK, pp.42-45.
Dodkins J. (1994). *ISO 9000 Quality Assessor Training Manual*, John Dodkins Associate, UK.
DTI (1990). *The Single Market: Testing and Certification*, Dept. of Trade & Industry, UK.
DTI (1991, Oct). *Managing in the '90s: The Quality Gurus*, Dept. of Trade & Industry, UK.
DTI (1992, Mar). *Managing in the '90s: Quality Circles*, Dept. of Trade & Industry, UK.
DTI (1992, Mar). *Managing in the '90s: Statistical Process Control*, Dept. of Trade & Industry, UK.
DTI (1994, Mar). *ISO 9000 A Positive Contribution to Better Business*, Dept. of Trade & Industry, UK.
DTI (1994, Nov). *Best Practice Benchmarking*, Dept. of Trade & Industry, UK.
Eicher L.D. (1992, June). "Quality Management in the '90s: The ISO 9000 Phenomena", *Quality Forum, Vol.18, No.2*, Institute of Quality Assurance, UK.
Feigenbaum, A.V. (1991). *Total Quality Control*, 3rd Ed., N.Y. McGrawHill Inc. pp.5-6, pp.11-14.
George S. and Weimerkirch A. (1994). *Total Quality Management: Strategies and Techniques Proven at Today's Most Successful Companies*, John Wiley & Sons, USA.
Hakes C. (1991). *Total Quality Management: A Key to Business Improvement*, Chapman & Hall, UK.
Hakes C. (1995). *The Self Assessment Handbook: For measuring corporate excellence*, Chapman & Hall, UK.
Henderson G.L. (1992, June). "Girobank: The first bank to win a British Quality Award", *Quality Forum, Vol.18, No.2*, Institute of Quality Assurance, UK.
Herzberg F. (1966). *Work and the Nature of Man*, World Publishing Co., USA.
Ho S.K. (1988). *Information Technology Development for Small & Medium Enterprises in Asian NICs and Japan*, Asian Productivity Organisation Press, Japan.
Ho S.K. (1993). "Problem Solving in Manufacturing", *Management Decision, Vol.31, No.7*, UK.
Ho S.K. (1993). "Transplanting Japanese Management Techniques", *Long Range Planning, Vol.26, No.4*, Pergamon Press, UK, pp.81-89.
Ho S.K. (1993, Aug). "If Japan Can, Why Can't We?", *Business Studies, Vol.5, No.5*, UK.
Ho S.K. (1994, Feb 23). "Proposal to Prevent Collapse of High-rises", *Star Newspaper*, Malaysia.
Ho S.K. (1994, Mar). "ISO 9000: the Route to Total Quality Management", *Quality World*, Institute of Quality Assurance, UK, pp.157-163.

Ho S.K. (1994). "Japanese Management A-Z", *International Review of Strategic Management, Vol.5, a Chapter*, UK, pp.171-186.

Ho S.K. (1994). "Is the ISO 9000 Series for TQM?", *Int. Journal of Quality & Reliability Management, Vol.11 No.9*, MCB, UK, pp.74-89.

Ho S.K. & Cho W. K. (1995, Jan). "Lessons for Industry from the Fast Food Business", *Management Services Journal*, Carfax Publishing, UK, pp.14-17.

Ho S.K. & Cicmil S. (1995). "TQM Transfer to SMI in Malaysia by SIRIM", *Total Quality Management, Vol.6, No.3*, Carfax Publishing, UK.

Ho S.K. & Cicmil S. (1995). "Can the S-S Problem Solving Method help to win the World Cup?", *Management Decision, Vol.33, No.7*, UK.

Ho S.K. & Fung C. (1995). "A Model of Excellence for TQM: LETQMEX", *Journal of Strategic Change, Vol.3, 165.1-12*, J. Wiley & Sons, UK.

Ho S.K. & Fung C. (1994, Dec). "Developing a TQM Excellence Model: Part 1", *TQM Magazine, Vol.6. No.6*, MCB, UK, pp.24-30.

Ho S.K. & Fung C. (1995, Feb). "Developing a TQM Excellence Model: Part 2", *TQM Magazine, MCB, Vo.7, No.1*, UK, pp.24-32.

Ho S.K. & Hamzah A. (1994, Aug.). "TQM Training for Small and Medium Industries in Malaysia", *Training for Quality Journal, Vol.2, No.2*, UK.

Ho S.K. & Lee N.C. (1994, Oct). "ISO 9000: A Strategic Quality Tool for the Construction Industry", *Campus Construction, Journal of the Chartered Institute of Building, Vol.2, Iss.3*, pp.13-15.

Ho S.K. & Wearn K.(1995). "A Higher Education TQM Excellence Model: HETQMEX", *Quality Assurance in Education Journal, Vol.3, No.3*, MCB, UK, pp.42-49.

Hutton M. (1995). "Successfully combining TQM with business process re-engineering", *Proceedings of the 1st World Congress for TQM*, Chapman & Hall, UK, pp.161-165.

Irvine G. (1991, July). "Systems for Managing Quality", *Computing & Control Engineering Journal*, pp.170-173.

Ishikawa K. (1986). *Guide to Quality Control*, Asian Productivity Organisation, Tokyo.

ISO (1994). *ISO 8402: Quality Vocabulary*, International Organization for Standardization, Geneva.

ISO (1994). *ISO 9000: 1994 Quality Management Standard*, International Organization for Standardization, Geneva.

ISO (1993). *ISO 9004-4:1993 Guidelines for Total Quality Management*, International Organization for Standardization, Geneva.

Juran J.M. (1988). *Juran on Planning for Quality*, The Free Press, New York.

Juran J.M. and Gryna F.M. (1988). *Quality Planning and Analysis*, New Delhi, Tata McGraw-Hill, pp.126-128.

Johnson P.L. (1993). *ISO 9000: Meeting the New International Standards*, McGraw-Hill Inc., USA.

Kepner C. and Tregoe B. (1965). *The Rational Manager*, McGraw-Hill, New York.

Kondo Y. (1989). *Human Motivation - A Key Factor for management*, 3A Corporation, Tokyo, pp.65-66

Kondo Y. (1995). "Quality and People", *Proceedings of the 1st World Congress for TQM*, Chapman & Hall, UK, pp.11-25.

Levitt T. (1981). "Marketing Intangible Products and Product Intangibles", *Harvard Business Review, May-June*, USA, pp.94-102.

Lip E. (1989). *The Chinese Art of Survival*, EPB Publishers, Singapore.

MacDonald J. & Piggott J. (1990). *Global Quality: the New Management Culture*, Mercury Publisher, USA.

MacGregor D. (1960). *The Human Side of Enterprise*, McGraw-Hill., USA.

Masaaki I. (1986). *Kaizen*, Random House, New York.

Matsushita K. (1988). *Quest for Prosperity: The life of a Japanese Industrialist*, PHP Institute Inc., Tokyo.

Mauro N.J. (1992). *A Perspective on Dr. Deming's Theory of Profound Knowledge*, British Deming Association, UK.

McBride P.S. (1993, June). "Shorts Total Quality Programme: Quality Management in Practice", *Quality Forum, Vol.19, No.2*, Institute of Quality Assurance, UK.

Melville S. and Murphy A. (1989). "Quality Improvements through People", *Quality Assurance, Vol.15, No.1*, UK, pp.25-28.

Migliore R.H. (1992). *Common Sense Management*, S. Abdul Majeed & Co., Malaysia.

Neave H.R. (1990). *The Deming Dimension*, SPC Press, Knoxville, Tennessee, p.300.

Neave H.R. (1995). "W. Edwards Deming (1900-1993): the man and his message", *Proceedings of the 1st World Congress for TQM*, Chapman & Hall, UK, pp.50-9.

NPB (1984). *Quality Control and QCCs - A report on a study trip to Japan*, National Productivity Board, Singapore.

Oakland J.S. (1989). *Total Quality Management*, Butterworth-Heinemann Ltd., Oxford.

Oakland J.S. (1995). "Business Process Re-engineering", *Proceedings of the 1st World Congress for TQM*, Chapman & Hall, UK, pp.95-108.

Osada T. (1991). *The 5-S: Five Keys to a Total Quality Environment*, Asian Productivity Organization, Tokyo.

Ouchi W.G. (1981). *Theory Z: How American Business Can Meet the Japanese Challenge*, Avon Books, New York.

Ovenden T.R. (1994). "Business Process Re-engineering: Definitely Worth Considering", *Total Quality Management Magazine, Vol.6, No.3*, MCB, UK, pp.56-62.

Parasuraman A., Zeithaml V. & Berry L. (1990). *Delivering quality service*, Free Press, USA.

Peratec (1994). *Total Quality Management: The key to business improvement*, 2nd Ed., Chapman & Hall.

Peters T. (1988). *Thriving on Chaos*, MacMillan, USA.

Pope N.W. (1979, Sep. 12). "More Mickey Mouse Marketing," *American Banker.*
Quality World (1995, Jan). *70,517 ISO 9000 Certifications Worldwide*, Journal of the Institution of Quality Assurance, p.37, UK.
Rees R.G. (1994) *Essential Statistics,* Chapman & Hall, UK.
Ross J.E.. (1993) *TQM: Text, Cases and Readings*, St. Lucie Press, USA.
Sallis E. (1993). *Total Quality Management in Education,* Kogan Page, UK.
Sayle A.J. (1988). "ISO 9000 -- Progression or Regression", *Quality Assurance News, Vol.14, No.2,* UK, pp.50-53.
Senju S. (1992). *TQC and TPM*, Asian Productivity Organization, Tokyo.
Shewhart W.A. (1933). *Economic Control of Manufactured Product,* D. Van Nostrand Company Inc., New York.
Smith S. (1988). "Perspectives: Trends in TQM", *The TQM Magazine, Vol.1, Iss.1,* UK, pp.5.
Starkey M.W. (1995). "Measurement with meaning: Using SPC to improve marketing performance", *Journal of Targeting, Measurement and Analysis for Marketing,* Henry Stewart Publisher, UK.
Straw D. (1988, May). "The Benefits of Independent Quality Assurance Registration to Buyers of Finished Components", *Transaction Institute of Metal Finishing, Vol.66,* pp.73-74.
Taguchi G. (1986). *Introduction to Quality Engineering,* Asian Productivity Organisation, Tokyo.
Vroman H.W. and Luchsinger V. (1994). *Managing Organisation Quality,* Richard Irwin, USA, pp.188-196.
Walton M. (1991). *Deming Management at Work,* Perigee Books, USA.
Wayne S.R. (1983, October). "Quality Control Circle and Company Wide Quality Control", *Quality Progress,* pp.14-17.
West A. and Phillips T.M. (1988, December). "Let's go with 5750, Experience of Implementing BS-5750 in a Small Chemical Company", *Symposium on Quality Assurance in the Process Industries,* Institution of Mechanical Engineers, Manchester.
Wheeler D.J. and Chambers D.S. (1992). *Understanding Statistical Process Control,* SPC Press Inc., Knoxville, Tenn., USA.
Whittington D. (1988). "Some Attitudes to ISO 9000: A Study", *International Journal of Quality and Reliability Management, Vol.6, No.3,* UK, pp.54-58.
Witcher B.J. (1990, Winter). "Total Marketing: Total Quality and the Marketing Concept", *The Quarterly Review of Marketing.*
Wong A.S. (1995). "Total Quality in Purchasing", *Proceedings of the 1st World Congress for TQM,* Chapman & Hall, UK, pp.181-184.
Wong S. (1995, Jan). "Business Process Re-Engineering or Continuous Improvement?", *Quality World,* Institute of Quality Assurance, UK.
Yip K K (1986). *Value Engineering,* Tsui's Publishing, Taiwan.

Suggested Answers to Self-Assessment Exercises

1.1 -- Quality Awareness Checklist

1	2	3	4	5	6	7	8	9	10	11	12	13	14	15	16	17	18
T	T	T	T	F	F	T	T	F	T	F	T	T	T	F	T	T	F

Excellent 16-18; Good 13-15; Average 10-12; Do it again <10.
(Remember, how many right answers you have achieved is less important than knowing the correct answers to all them by now!)

1.2 -- TQM Awareness Checklist

1	2	3	4	5	6	7	8	9	10	11	12	13	14	15	16	17	18	19	20	21	22	23	24
T	F	F	T	T	F	T	F	F	T	T	F	T	T	F	T	F	F	T	T	F	F	F	T

Excellent 21-24; Good 17-20; Average 13-16; Do it again <12.
(Remember, how many right answers you have achieved is less important than knowing the correct answers to all them by now!)

1.3 -- The F-Test

6-F's.
You are very lucky if you can get it right the first time! This demonstrates human fallibility in a very elementary inspection operation. The message is that if you want to provide quality products or services to your customers, both internal and external, you should double-check your work first before you pass it on. This message is expressed in a more subtle way by the 1-10-100 Rule that follows. Try the F-test with your colleagues and comment on the results!

3.3 -- Rating Your Communication

1. Feedback! Feedback! Feedback! Most of us do not give or get enough feedback to complete the communication loop. Feedback can be written, verbal, and non-verbal.
2. You can ask, "What is your reaction to my suggestion (or statement)?" Also look carefully for non-verbal signals as they tell you more than words do. When writing, ask for a written response.
3. Ask them how you can help them. Listen attentively. Provide verbal and non-verbal feedback. Be open-minded.
4. Same as No.3, and **Know Your Purpose.**
5. Self-evaluation and feedback from others. Do I feel more competent as a communicator? Are my projects completed in less time with less errors and stress?
6. Communication is the Number One way to create quality awareness. Write the word 'QUALITY' on the walls and floormats, and write about it often in your organisation's newsletter. Put quality statistics and improvement charts in the dining and lounge areas. Spread the word and say it with pride. Make quality an integral part of your company culture. Management should talk about quality regularly. Make it the first agenda item at department meetings.

General Remarks to Suggested Solutions to Chapter 7

As a general remark to suggested answers, it is important to emphasise the fact that sometimes the interpretation of ISO 9001: 1994 could be quite flexible. Therefore, the attempt here is to give the perception of the author which may not be definitive.

7.1 -- Interpretation of ISO 9001

1. If the customer does not agree.
2. If the capability is relevant to the quality of the output which are part of the customer's requirement.
3. Clause 4.15.3.
4. Customer verification of subcontracted product (Clause 4.6.4.2).
5. (a)   4.9 (f)
   (b)   4.9 (a)
6. Clause 4.10.5.
7. As a result of quality improvement.
8. Refer to Clause 4.4.2; Clause 4.4.3; Clause 4.4.7.
9. Refer to Clause 4.4.6.
10. 20 procedures in response to the 20 clauses.

11. Some service industries (e.g. management consultants)
12. When they are signed by authorised person and dated.
13. Through inspection, quality audit, documentation and traceability.
14. (a)   Clause 4.6.4.
    (b)   Clause 4.6.2.
15. When they have effects on quality.
16. Clause 4.3.3.

7.2 -- ISO 9001 Quality Audit

1. Clause 4.5.2 (b).
2. Clause 4.15.3.
3. Clause 4.8 & 4.18
4. Clause 4.17
5. Clause 4.3 & 4.18
6. Clause 4.6.3 & 4.10.1
7. Clause 4.11.2 (g) & (i)
8. Clause 4.3 & 4.5.3
9. Clause 4.11
10. Clause 4.10.3 & 4.20
11. Clause 4.9 (c)
12. Clause 4.16

7.3 -- ISO 9001 Mock Audit

1. Clause 4.1.2.1; 4.6.4.2 & 4.9 (c)
2. Clause 4.17
3. Clause 4.5.2 (a)
4. Clause 4.11.1
5. Clause 4.9 (d) & 4.10.3
6. Clause 4.10.3
7. Clause 4.11.1
8. Clause 4.16
9. Clause 4.11.2
10. Clause 4.5 & 4.11
11. Clause 4.10.4
12. Clause 4.5.3; 4.9 & 4.15.4
13. Clause 4.10.5

SOURCES OF INFORMATION AND ADVICE

In the UK:

British Deming Association (BDA)

The BDA was formed in 1987 by organisations and individuals who had gained fresh insight into their approach to management through the work of Dr W Edwards Deming. Dr Deming became its Honorary Life President in 1988. The BDA exists to spread awareness of the Deming management philosophy and to help members put it into practice. Members of the BDA have the opportunity to participate in the work of their Research Groups, regional activities, and other initiatives. The BDA is a not-for-profit organisation. As such, it is not involved in consultancy, although it does lead an education programme for an Alliance of consultants who desire to centre their work on the Deming philosophy.
Enquiries may be addressed to:
The British Deming Association,
The Old George Brewery, Rollestone Street, Salisbury, Wiltshire SP1 1DX, UK.
Tel: +44 1722 412138   Fax: +44 1722 414428

British Quality Foundation (BQF)

The BQF exists to enhance the performance and effectiveness of all types of enterprise and organisation within the UK through the promotion of TQM practices. Membership is open to all industrial, commercial and other corporate organisations. Through seminars, conferences, publications and other services and activities the BQF aims to promote the use and understanding of TQ as a basis for achieving improvements in performance relevant to the needs of all enterprises and organisations. The BQF also runs the UK Quality Award.
Enquiries may be addressed to:
The British Quality Foundation,
Vigilant House, 120 Wilton Road, London SW1V 1JZ, UK.
Tel: +44 171 931 0607   Fax: +44 171 233 7034

British Standards Institution (BSI)

BSI is responsible for preparing British Standards which are used in all industries and technologies. It also represents British interests at international standards discussions to ensure that European and world-wide standards will be acceptable to British industry. Information about all aspects of BSI's work including sales information on British, international and foreign national standards can be obtained from:
British Standards Institution,
389 Chiswick High Road, London W4 4AL, UK.
Tel: +44 181 996 9000   Fax: +44 181 996 7400

Department of Trade and Industry (DTI)

The DTI's Enterprise Initiative offers businesses a co-ordinated programme of information, advice and assisted consultancies on key areas of quality management practice. The DTI QA Register lists organisations in the UK and abroad whose quality systems have been assessed to BS EN ISO 9000 by an accredited UK certification body. The 1995 edition includes around 35,000 entries. General enquiries in connection with DTI's efforts to promote TQM may be addressed to:
Department of Trade and Industry,
Room 5.77, 151 Buckingham Palace Road, London SW1W 9SS, UK.
Tel: +44 171 215 5847   Free Phone (UK only): (0800) 500 200

Institute of Quality Assurance (IQA)

The IQA, established in 1916, has the largest membership amongst professional bodies involved with Quality in Europe, and represents the UK in the European Organisation for Quality. It produces the monthly magazines *Quality World* and *Quality Forum*, and provides training courses, conferences and seminars. The IQA's National Quality Information Centre provides information, books, journals and a useful database to the business community, particularly smaller businesses, and publishes a list of training aids available in the UK. The IQA also administers the International Register of Certification Auditors (IRCA). The scheme controls the registration of auditors (assessors) and registers training organisations that provide related courses. Individual and company membership is available and enquiries should be addressed to:
The Institute of Quality Assurance,
PO Box 712, 61 Southwark Street, London SE1 1SB, UK.
Tel: +44 171 401 7227   Fax: +44 171 401 2725

National Accreditation Council for Certification Bodies (NACCB)

The NACCB was launched in 1984 to make recommendations to the Secretary of State for Trade and Industry concerning the competence and impartiality of certification bodies, leading to their accreditation by the Secretary of State. Accreditation enables certification bodies to display the National Accreditation Mark alongside their own certification marks. Enquiries may be addressed to:
The Secretary (NACCB),
Audley House, 13 Palace Street, London SW1E 5HS, UK.
Tel: +44 171 233 7111   Fax: +44 171 233 5115

National Society for Quality through Teamwork (NSQT)

The NSQT aims to enable UK companies to continuously improve business performance through the effective use of teamwork. To achieve this aim it utilises the experience and energy of its member organisations, currently well over 500, within which a whole variety of teamwork is employed. The NSQT provides a network of regional meetings throughout the UK, seminars, workshops, and conferences for the exchange of experience. It also runs training courses, management facilitation, and an advisory service. The NSQT maintains links to the rest of Europe as a founder member of EFQCA (European Foundation for Quality Circles and Management Associations), and to the USA through Association with AQP (Association for Quality and Participation). For more information contact:
National Society for Quality through Teamwork,
2 Castle Street, Salisbury, Wiltshire SP1 1BB, UK.
Tel: +44 1722 326 667   Fax: +44 1722 331313

In Europe:

The European Foundation for Quality Management (EFQM)

The EFQM's mission is to support the management of Western European companies in accelerating the process of making quality a decisive influence for achieving global competitive advantage and to stimulate and, where necessary, assist all segments of the Western European community to participate in quality improvement activities and to enhance the quality culture. The EFQM has a membership of nearly 400 leading European businesses, all of whom recognise the role of quality on achieving competitive advantage. The European Quality Award, launched in 1991, is a key focus of the organisation. Enquiries may be addressed to:

European Foundation for Quality Management,
Avenue des Pleiades 19, 1200 Brussels, BELGIUM.
Tel: +32 2 775 35 11   Fax: +32 2 779 12 37

In the USA:

US National Institute of Standards and Technology (NIST)

NIST is an agency of the Department of Commerce's Technology Administration. Its goals are: to aid US industry through research and services; to contribute to public health, safety and the environment; and to support the US scientific and engineering research communities. Much of NIST's work relates directly to quality and to quality-related requirements in technology development and technology utilisation. NIST also manages 'The Foundation for the Malcolm Baldrige National Quality Award (MBNQA)'. Contact address is:

National Institute of Standards & Technology,
Route 270 and Quince Orchard Road, Administration Building Room A537, Gaithersburg, MD 20899-0001, USA.
Tel: +1 301 975 2036   Fax: +1 301 948 3716

American Society for Quality Control (ASQC)

ASQC is dedicated to facilitating continuous improvement and increased customer satisfaction by identifying, communicating and promoting the use of quality principles, concepts and technologies. ASQC strives to be recognised throughout the world as the leading authority on, and champion for, quality. It assists in administering the USA MBNQA programme under contract to NIST. Contact Address is:

American Society for Quality Control,
P O Box 3005, Milwaukee, WI 53210-3005, USA.
Tel: +1 414 272 8575   Fax: +1 414 272 1734

In the 'Rest of the World':

Sources of help from own country normally include your:

- National Productivity Organisation
- National Ministry/Department of Trade and Industries (normally manages the National Quality Award Scheme as well)
- National Standards and Industrial Research Institution
- National Quality Accreditation Body (for ISO 9000)
- National Quality Professional Institution

Additional Source: Asian Productivity Organization (APO)

APO is an Asian-Pacific organisation consisting of 16 membership countries. The aim is to promote the productivity and quality of the goods and services within its membership countries. It also publishes productivity and quality management books which are disseminated world-wide. Many of the publications are based on the experience of its founder country, Japan. APO is widely recognised as one of the most important organisations in the promotion of TQM.

Asian Productivity Organization,
4-14 Akasaka 8-Chome, Minato-ku, Tokyo, JAPAN.
Tel: +81 33 408 7221   Fax: +81 33 408 7220

INDEX

4Cs of TQM 51
5-S
    audit worksheet 92-5
    cleaning 72-3
    daily activity for the office 97
    daily activity for the workshop 96
    definition 47
    discipline 75-8
    elements of 63
    evaluation 88
    implement 84-91
    implementation plan 98
    in the office 79-82
    neatness 68-71
    organisation 64-8
    promotional campaign 85-7
    standardisation 74-5
    training 87
    use of 82-4
7-Eleven 154

Allied Quality Assurance Publication (AQAP) 175
American Society for Quality Control (ASQC) 286
ANOVA 251, 253-4
Asian Productivity Organization (APO) 1, 288

Bajaria, Hans J. 155
benchmarking 12
best practice benchmarking (BPB)
    making it work 114-17
    meaning of 114
Boots The Chemists 130
Bowater Containers Southern 198
breakdowns 84
British Deming Association 27, 284
British (see UK) Quality Award 236-7
British Quality Foundation (BQF) 235, 284
British Standards Institution (BSI) 175, 195, 285
British Telecom 119
business process re-engineering (BPR)
    definition 47
    dilemma 101-2
    implement 131
    meaning of 99-101
    top-down approach 132
    use of 130
    velocity ratio 101

Co-op Bank 133
commitment 51-2, 60
communication 53-4, 61
company wide quality control (CWQC) 3, 35
competence 52-3, 60
competitive advantage 154
continuous improvement (see kaizen)
control charts
    c-chart 152
    for attributes 150-2
    for variables 146-9
    lower control limit (LCL) 146-9
    mean chart 146, 149
    p-chart 150-1
    process improvement 152
    for QCC 154
    range chart 147, 149
    upper control limit (UCL) 146-9
corporate strategy 6
corporate vision 8
Crosby, Philip
    background 5, 32, 41
    do it right first time 32
    zero defects 32-3
customer satisfaction 108

Deming, W. Edwards
   14 Points 29
   background 2, 22, 40
   Cycle 22, 55-6
   Doctor's Orders 27
   new climate 23
   Prize 38, 226-7
   system of profound knowledge 25
Department of Trade and Industry (DTI) 187, 285
Dow Chemicals 8
DTI Import Licensing Branch 198

efficiency 83-4
employee attitude survey 197
employee empowerment 51, 171
European Foundation for Quality Management (EFQM) 232-3, 286
European Quality Award (EQA)
   background 232
   checklist 245-8
   framework 233
   results 234
extraordinary customer satisfaction 5

F-Test 19
fool-proofing (see Poka-Yoke)
four pillars of TQM 50
Fujikoshi 218

Girobank 235

Hydrapower Dynamics Ltd. 168

industrial cycle 11
information technology (IT) 259
Institute of Quality Assurance (IQA) 285
International Organization for Standardization (ISO) 47, 175
Ishikawa, Kaoru 35-36, 41, 154
ISO 9000 (ISO 9001)
   20 clauses 213-14
   benefits of 195-6
   content of 177-8
   definition 48
   family 210
   four pillars 20
   implement 188-95
   interpretation 201
   meaning of 175

   need of 186-7
   purpose 176-7
   win-win situation 178-9
   world share of certifications 176
ISO 9004-4: 1993
   background 12
   definition 48
   outline 243-4
   quality improvement 226

Japan Institute of Plant Engineers (JIPE) 48, 215
Japan Union of Scientists and Engineers (JUSE) 2, 22, 136
Japanese
   economic miracles 1
   quality control 2
   QCC experience 136
   quality miracle 1
John Laing plc 9
Johnson Electronic Co. Ltd. 166
Juran, Joseph M.
   background 2, 8, 30, 41
   last word conference 32
   quality spiral 31
   quality trilogy 30-1
just in time (JIT) system 36-7

kaizen (continuous improvement)
   dilemma 101-2
   in ISO 9004-4 12
   meaning of 55-6
   practices 57-8
   strategy 56-7
keep it short and simple (KISS) 44
Kepner-Tregoe 156
Komatsu Company 56
Kondo, Yoshio 38-41

leadership 24, 51, 229, 245
Leicester Business School 9
Lucas Industries 124
Lyondell Petrochemical 172

Malaysian Government 44
Malcolm Baldrige National Quality Award (MBNQA)
   1991 Winner 231
   background 228
   examination criteria 229-30

management representative 189
Matsushita Electric 8
MBI 199
mission statements 8

National Accreditation Council for
    Certification Bodies (NACCB) 286
National Broadcasting Corporation (NBC)
    3, 31
National Society for Quality through
    Teamwork (NSQT) 286
Neave, Henry 22, 27, 40
need 65-6

one-is-best 66, 80-2

Parasuraman, A. 110
PDCA (see Deming Cycle)
PDSA (see Deming Cycle)
Peters, Tom 120
PHP Institute Inc. 269
PM Excellence Award 218
Poka-Yoke 36
Praxis Software Engineering Company
    199
problem solving 155-6

quality
    action team 189
    chain 11
    consciousness checklist 18
    control 2
    definition 5
    environment 84
    improvement 12
    problems in industry 14-6
    steering committee 189
quality audit
    definition 181
    exercise 202-3
    facilitate 182-6
    hints 186
    mock 204-9
    need 186-8
    types 181-2
quality control circles (QCC)
    establish 169
    in-house presentation 137
    Japanese experience 136
    meaning of 135

need for 168
suggestion schemes 137
quality control tools
    cause-and-effect diagram 144
    check sheet 139-40
    control charts 146-54
    definition 47
    graphs 140-2
    meaning of 138
    Pareto analysis 142-3
    problem solving 165
    process flowchart 138-9
    scatter diagram 145-6
    specifications 173
    use 154
quality cost
    analysis 4
    appraisal costs 34
    failure costs 34
    prevention costs 33
quality gurus (see TQM gurus)
quantification 75
quotable standards 175

Rank Xerox's 117
Ricardo, David 1
Rover Group 238

safety 83
satisfying customers 51
Scandinavian Air System 107
school of quality 112-13
Sea-Land 8
service industry 15, 104, 106
service quality 104
Servqual
    applications in education 110-1
    dimensions 110
    gaps 110-1
    Model 109
Shewhart, W.A. 22
Shingo, Shigeo 36-8, 41
Short Brothers 236-7
SIRIM 44, 89
S-S Method of Problem Solving
    S-S Method 156-66
    S-S Method Flowchart 157
    S-S Method Map 159-60
    S-S Method Worksheet 159-60
Starkey, M. 153-4

strategy for total quality 9

TNT Express (UK) Ltd. 238
top management commitment 84-5
TQM
    4-Cs 52-9
    concepts 11-3
    consciousness checklist 17
    definition 4
    four pillars 50-2
    improvement tools 50-1
    ten commandments 50
TQM gurus
    ideas 40
    implement the ideas 41
    meaning of 21
    need of 13-5
TQM kitemarks
    achievement of 240-1
    meaning of 225
    use of 239
TQMEX Model
    content of 44-8
    impact on improvement 272
    implementation cycle 273-4
    implementation programme 273
    logic of 48-9
    need for 43
    pre-requisites 49
TQMEX validation
    background 249-50
    open-ended questions 259-60, 266-8
    questionnaire 250-1
    sampling 249-50
    significance of ISO 9000 270-3
    UK and Japanese companies survey 262-70
    UK companies survey 251-62
total productive maintenance (TPM)
    autonomous maintenance 215
    content of 215-20
    definition 49
    equipment maintenance 216
    implement 223
    meaning of 213-4
    rating 223-4
    use of 220-3
    work-time breakdown 221-2
total quality marketing (TQMar)
    areas 121-4
    corporate environment 120
    marketing oriented 118-9
    meaning of 118
    process 120-1
    TV Advertising 123
total quality purchasing (TQPur)
    performance measurement 126-9
    meaning of 125
Toyota Auto Body Co. 218
transparency 74-5
trouble maps 75

UK Quality Award 235-8
US National Institute of Standards and Technology (NIST) 287

value
    added analysis 103
    analysis 103
    engineering 103-4
    in service industry 106
visualising conditions 75
visual management 74

want 65-6
Wellex Corp. 90
World Cup 1990 Semi-Final 160-5

zero defects 32-3, 38, 173
Zytec Corp. 28, 231